Safety at Work and Emergency Control

Second Edition

Safety at Work and Emergency Control
A Holistic Approach

Second Edition

Benedito Cardella

CRC Press
Taylor & Francis Group
Boca Raton London New York

CRC Press is an imprint of the
Taylor & Francis Group, an **informa** business

This book was previously published in Portuguese, entitled "Segurança no Trabalho e Prevenção de Acidentes" - Uma Aboragem Holística" (Safety at Work and Accidents Prevention - A Holistic Approach) published by Editora Atlas LTDA - Gen | Grupo Editorial Nacional, Rio de Janeiro, Brasil, under the imprint of GEN/Atlas.

CRC Press
Taylor & Francis Group
6000 Broken Sound Parkway NW, Suite 300
Boca Raton, FL 33487-2742

© 2019 by Taylor & Francis Group, LLC
CRC Press is an imprint of Taylor & Francis Group, an Informa business

No claim to original U.S. Government works

Printed on acid-free paper

International Standard Book Number-13: 978-1-138-61540-3 (Hardback)
International Standard Book Number-13: 978-0-429-46312-9 (eBook)

This book contains information obtained from authentic and highly regarded sources. Reasonable efforts have been made to publish reliable data and information, but the author and publisher cannot assume responsibility for the validity of all materials or the consequences of their use. The authors and publishers have attempted to trace the copyright holders of all material reproduced in this publication and apologize to copyright holders if permission to publish in this form has not been obtained. If any copyright material has not been acknowledged please write and let us know so we may rectify in any future reprint.

Except as permitted under U.S. Copyright Law, no part of this book may be reprinted, reproduced, transmitted, or utilized in any form by any electronic, mechanical, or other means, now known or hereafter invented, including photocopying, microfilming, and recording, or in any information storage or retrieval system, without written permission from the publishers.

For permission to photocopy or use material electronically from this work, please access www.copyright.com (http://www.copyright.com/) or contact the Copyright Clearance Center, Inc. (CCC), 222 Rosewood Drive, Danvers, MA 01923, 978-750-8400. CCC is a not-for-profit organization that provides licenses and registration for a variety of users. For organizations that have been granted a photocopy license by the CCC, a separate system of payment has been arranged.

Trademark Notice: Product or corporate names may be trademarks or registered trademarks, and are used only for identification and explanation without intent to infringe.

Visit the Taylor & Francis Web site at
http://www.taylorandfrancis.com

and the CRC Press Web site at
http://www.crcpress.com

I dedicate this book to those who saw moments of joy, enthusiasm, and hope turning into sadness, dejection, and despair. I wish this book to contribute effectively to the reduction of accidents.

Contents

About the Book xv
Preface xvii
Acknowledgments xix
Author xxi
Introduction xxiii

Chapter 1 Holistic Approach 1

    1.1 Harmonizing the Paradigms 1
    1.2 Harmonizing Survival and Happiness 3
    1.3 Approaching Accidents 3
    1.4 Approaching Organizations 5
        1.4.1 Addressing Operational Systems 5
        1.4.2 Addressing Organizational Systems 6
        1.4.3 Addressing Activities 6
        1.4.4 Addressing the Life Cycles of the Organization, Products, and Personnel 6
        1.4.5 Making Diagnosis 7
        1.4.6 Assessing the Operational System 7
        1.4.7 Assessing the Organizational System 7
    1.5 Performing Interventions 8
    Bibliography 9

Chapter 2 Safety at Organizations 11

    2.1 Safety Function 11
    2.2 Organizations 11
    2.3 People Needs 13
    2.4 Adherence 14
        2.4.1 Adherence Client–Organization 15
        2.4.2 Adherence Organization–Client 16
        2.4.3 Adherence Organization–Organization 16
        2.4.4 Perceived and Real Adherence 16
        2.4.5 Adherence and Quality 16
        2.4.6 Adherence Control 16
    2.5 Leadership 18
        2.5.1 Leadership and Administrative Game 18
        2.5.2 Relation Leader–Subordinate 19
    2.6 Organizational Culture 20
    2.7 Management System 22
    2.8 Holistic Management System 23

		2.8.1	Basic Principles	23
		2.8.2	Policy	23
		2.8.3	Responsibility	24
		2.8.4	Scope	24
		2.8.5	Management Methodology	24
		2.8.6	Programs' Structure	24
	2.9	Organizational Climate		24
	2.10	Resources of the Organizations		25
	2.11	Organizational Field		27
	2.12	Organizational Development		29
	Bibliography			31

Chapter 3	Risk Management		33
	3.1	Risk Control	33
	3.2	Risk Management	33
	3.3	Principles	33
	3.4	Goals	33
	3.5	Policy	34
	3.6	Guidelines	34
	3.7	Strategies	34
	3.8	Organizational Systems	34
	3.9	Methodology	35
	3.10	Action Areas	35
	3.11	Life Cycle	35
	3.12	Programs	36
	3.13	Monitoring	36
	3.14	Indicators	37
	3.15	Audits	37
	3.16	Diagnosis	38
	Bibliography		38

Chapter 4	Emergency Management			39
	4.1	Emergency Control		39
	4.2	Emergency Control Principles		40
	4.3	Emergency Management Policy		40
	4.4	Emergency Management Guidelines		40
	4.5	Emergency Management Strategy		41
	4.6	Emergency Management Methodology		41
		4.6.1	Emergency Control Function Deployment	41
	4.7	Action Areas		41
		4.7.1	Emergency Action Plan	42
	4.8	Emergency Control Organization		42
		4.8.1	Mission	42
		4.8.2	Clients	42

		4.8.3	Organizational Structure	43
		4.8.4	Resources	43
		4.8.5	Action Groups	43
	Bibliography			49

Chapter 5 Risk Management in Interventions 51
    5.1 Concept of Intervention 51
    5.2 Types of Intervention 51
    5.3 Impact and Risk Control 52
    5.4 Interfaces 53
    5.5 Permissions and Licenses 54
    Bibliography 54

Chapter 6 Risk Analysis and Control 55
    6.1 Concept and Methodology 55
    6.2 Harm Generation Mechanism 55
        6.2.1 Aggressive Action on Targets 55
        6.2.2 Failure of Risk Control 56
    6.3 Hazard Identification 56
    6.4 Risk Assessment 57
        6.4.1 Risk Factors 57
        6.4.2 Frequency Assessment 57
        6.4.3 Consequence Assessment 60
        6.4.4 Semiquantitative and Qualitative Risk Assessment ... 61
        6.4.5 Risk Control Instruments 63
    6.5 Risk Control 64
        6.5.1 Process Control 64
        6.5.2 Risk Control Model 66
    6.6 Action Plan for Risk Control 68
    Bibliography 70

Chapter 7 Risk Analysis Techniques 71
    7.1 Preliminary Risk Analysis 71
        7.1.1 PRA—Object and Focus 71
        7.1.2 PRA—Method 71
        7.1.3 Auxiliary Techniques 71
        7.1.4 Complementary Techniques 72
        7.1.5 Formulary 72
    7.2 Hazard and Operability Studies 72
        7.2.1 Hazop's Object and Focus 72
        7.2.2 Hazop—Method 73
        7.2.3 Continuous Process 73
        7.2.4 Discontinuous Process 74

		7.2.5	Auxiliary Techniques 75
		7.2.6	Complementary Techniques 75
		7.2.7	Formulary 76
	7.3	Failure Modes and Effects Analysis 76
		7.3.1	FMEA—Object and Focus 76
		7.3.2	FMEA—Method 76
		7.3.3	Auxiliary Techniques 76
		7.3.4	Complementary Techniques 76
		7.3.5	Formulary 77
	7.4	What-If? 77
		7.4.1	What-If—Object and Focus 77
		7.4.2	What-If—Method 77
		7.4.3	Auxiliary Techniques 78
		7.4.4	Complementary Techniques 78
		7.4.5	Formulary 78
	7.5	Checklist 78
		7.5.1	Checklist—Object and Focus 78
		7.5.2	Checklist—Method 79
		7.5.3	Auxiliary Techniques 79
		7.5.4	Complementary Techniques 79
		7.5.5	Formulary 79
	7.6	Fault Tree Analysis 79
		7.6.1	FTA—Object and Focus 79
		7.6.2	FTA Method 79
		7.6.3	Auxiliary Techniques 80
		7.6.4	Complementary Techniques 80
		7.6.5	Formulary 80
	7.7	Event Tree Analysis 81
		7.7.1	ETA—Object and Focus 81
		7.7.2	ETA—Method 81
		7.7.3	Auxiliary Techniques 82
		7.7.4	Complementary Techniques 82
		7.7.5	Formulary 82
	7.8	Critical Incident Technique 82
		7.8.1	Critical Incident Technique—Object and Focus 82
		7.8.2	CIT—Method 83
		7.8.3	Auxiliary Techniques 83
		7.8.4	Complementary Techniques 83
		7.8.5	Formulary 83
	7.9	Comparative Analysis 83
		7.9.1	Comparative Analysis—Object and Focus 83
		7.9.2	Comparative Analysis—Method 83
		7.9.3	Auxiliary Techniques 83
		7.9.4	Complementary Techniques 83
		7.9.5	Formulary 84
	7.10	Interaction Matrix 84

Contents

    7.10.1 Interaction Matrix—Object and Focus 84
    7.10.2 Interaction Matrix—Method 84
    7.10.3 Auxiliary Techniques 85
    7.10.4 Complementary Techniques 85
    7.10.5 Formulary 85
  7.11 Planned Inspection 85
    7.11.1 Planned Inspection—Object and Focus 85
    7.11.2 Planned Inspection—Method 86
    7.11.3 Auxiliary Techniques 87
    7.11.4 Complementary Techniques 87
    7.11.5 Planned Inspection Formulary 87
  7.12 Abnormal Occurrences Analysis 87
    7.12.1 Abnormal Occurrences Analysis—Object and Focus 87
    7.12.2 AOA—Method 88
  7.13 Cause Tree Analysis 89
  Bibliography 91

Chapter 8 Value Analysis in Safety 93
  8.1 Integrating Value and Risk Analysis 93
  8.2 Aggressive Function 93
  8.3 Basic Concepts of Value Analysis 94
    8.3.1 Object of Value Analysis 94
    8.3.2 Function 94
    8.3.3 Quality 94
    8.3.4 Price 95
    8.3.5 Absolute Value 95
    8.3.6 Relative Value 95
    8.3.7 Clients of the Organization 95
  8.4 Functions Classification 96
    8.4.1 Basic—Complementary and Supplementary Functions 96
    8.4.2 Auxiliary Function 97
    8.4.3 Use Function 97
    8.4.4 Esteem Function 97
    8.4.5 Necessary Function 97
    8.4.6 Unnecessary Function 97
    8.4.7 Support and Collateral Functions 97
    8.4.8 Aggressive Function 98
    8.4.9 Real Function 98
    8.4.10 Fictitious or Imaginary Function 98
    8.4.11 Perceived Function 98
    8.4.12 Revealed and Undisclosed Functions 98
    8.4.13 Vital Function 99
    8.4.14 Passive and Active Functions 99

	8.5	Functional Diagram	99
	8.6	Comparing Functions	101
	8.7	Holistic Diagram	101
	8.8	Questioning Functions	102
	Bibliography	104
Chapter 9	Human Failures	105
	9.1	Failure Analysis	105
	9.2	Technical Failure	107
	9.3	Careless Failure	108
		9.3.1 Carelessness Failure Characterization	108
		9.3.2 Inattention and Confusion	109
		9.3.3 Handling Carelessness Failures	109
		9.3.4 "Inadvertent" Equipment Failures	110
	9.4	Conscious Failure	110
		9.4.1 Conscious Failure Characterization	110
		9.4.2 Conscious Failure Mechanism	112
		9.4.3 Attitude	112
		9.4.4 Posture	114
		9.4.5 Behavior's Consequences	114
		9.4.6 Safe Behavior's Driving Force	116
		9.4.7 Behavior Control	118
		9.4.8 Leadership and Behavior	120
		9.4.9 Cultural Change	120
		9.4.10 Conscious Failure Treatment	121
	9.5	Compound Failure	121
	9.6	Failure Promotors	121
		9.6.1 Primary Agent	121
		9.6.2 Secondary Agent	122
		9.6.3 Command Agent	123
		9.6.4 Intruder Agent	124
	9.7	Common Cause Failure	124
	Bibliography	125
Chapter 10	Safety Concepts	127
	10.1	Object of Study	128
	10.2	Aggregate	128
	10.3	System	128
	10.4	Normal State	129
	10.5	Abnormal State	130
	10.6	Loss	131
	10.7	Abnormal Occurrences	131
	10.8	Impactful Agent and Target	132
	10.9	Harmful Agent and Susceptibility	133

Contents

10.10 Aggressive Agent and Vulnerability 134
10.11 Sources of Aggressive Agents 136
10.12 Aggressive Action Field 137
10.13 Exposure 138
10.14 Alarm 139
10.15 Promoters and Inhibitors 139
10.16 Containment and Retention 140
10.17 Rupture Agent 141
10.18 Isolation 141
10.19 Restoration 141
10.20 Combat 142
10.21 Protection 142
10.22 Defense 143
10.23 Recovery Systems 143
10.24 Hazard and Quality 144
10.25 Hazardous Event 145
10.26 Harmful and Damaging Events 146
10.27 Emergency 146
10.28 Accident—Incident—Near Miss 147
10.29 Risk 147
10.30 Maintenance 148
10.31 Operation 149
10.32 Project 150
10.33 Safety 150
Bibliography 151

Index 153

About the Book

This book treats "the accident" as a multifaceted phenomenon, resulting from complex interactions between physical, biological, psychological, cultural, and social factors. Addressing safety with holistic vision, it combines two complementary approaches: the reductionist, to study the factors in detail, and the systemic, to understand how they interrelate. It includes 33 concepts that provide a clear and logical understanding of every factor involved in any activity or situation regarding safety.

The author developed concepts and methods to boost safety performance. Organizational field, adherence, and administrative game explain why things happen or do not happen in the organizations. The aggressive function integrates value analysis and risk analysis. An individual adopts a safe or unsafe behavior the same way he decides to buy one product or another. Safety is a function placed at the same importance as its sisters: productivity, quality, environmental preservation, and human development. Risk is a process variable and as such one can control it.

Preface

Three reasons led me to write this book.

The first reason comes from knowledge. Working 27 years in a Petrobras refinery in Brazil, as process engineer, shift coordinator, head of operations, and health, safety, and environment manager, I learned about accident prevention and emergency control. Besides teaching me practical subjects, the refinery gave me opportunities to attend courses and seminars. All courses were helpful. The good ones, by teaching me; and the bad ones, by awakening the interest in developing something better. However, outside of the workplace, at home and on the streets, more remarkable things happen about safety than in industrial units. In addition, safety knowledge is also in the Bible, poems, and subjects such as history, anthropology, and psychology, not to mention physics, chemistry, and mathematics.

The second reason comes from my will. My early memory of my brother's death as a result of a domestic accident developed my motivation for safety. The motivation for environmental preservation resulted from having learned to take care of plants and animals.

The third reason comes from duty. Every accident with people close or distant, including those physically distant, but known because of newspaper, radio, and television stories, revived the feeling that I should do something to prevent their occurrence. It is not hard to build an unbearable list of names, some are well-known, such as Ayrton Senna, Christopher Reeve (Superman), Princess Diana, and John Denver; others are people from the readers' lives, including family and friends; and the unknowns of the vast majority, included in cold statistics. Three reasons, one goal: to contribute to the reduction of one of the main causes of human unhappiness.

Since the first edition in 1999, safety continued to guide my actions, spreading concepts, techniques, and experiences; publishing technical papers; providing consultancy services; and training more than 2,000 employees from different companies. Acting as safety expert for the Labor Justice, I went beyond the industrial field, evaluating working environments for airports and hospitals, among others.

Dreaming of new experiences in other cultures, I traveled to Kuwait in 2006. Living in the Middle East, I worked in the United Arab Emirates, Oman, Saudi Arabia, Qatar, Egypt, Tunisia, Thailand, Kazakhstan, and Italy. In addition, I have been working with people from Greece, Australia, and many countries in Asia for almost 12 years.

My new experiences led me to develop a second edition with more details on safety in organizations and human failures.

I wrote it in Portuguese, but I would like to share its contents with English speakers. That was a new challenge because it demanded more than a simple translation. Those who know the Portuguese editions will note the differences. For example, the chapter on emergency control starts with a quote from Sir Winston Churchill: "Plans are of little importance, but planning is essential."

Acknowledgments

Thank you very much,
To my parents, for putting me on the way to school;
To Petrobras, for the opportunities of development;
To people who contributed ideas, criticisms, and suggestions, encouraging the development of this work.

Author

Benedito Cardella is a chemical and safety engineer. He was a process engineer and operations and health, safety, and environment (HSE) manager of a Petrobras Refinery where he promoted an integrated system of safety, environment, and occupational health, and designed and developed the Emergency Control Organization. He has been acting as an HSE consultant in Brazil and several Asian countries.

Introduction

> The prudent see danger and take refuge, but the simple keep going and pay the penalty.
>
> *Proverbs: 22:3*

Every science has an object. The object is the sector of reality that the science studies. Consider physics. This science deals with phenomena that do not alter the nature of substances. Force, velocity, acceleration, density, and mass are concepts in physics used to understand physical phenomena and solve practical problems.

Accident prevention demands the study of phenomena that cause damages and losses to people, properties, and the environment. The science of accident prevention and emergency control should have those phenomena as objects and would be a multidisciplinary science to approach safety at work, at home, in public spaces, and in the environment. Analogously to what physics does, it would define basic concepts such as damage, loss, danger, risk, aggressive agent, aggressive energy, containment, protection, and emergency, among others. With them, one would study accidental phenomena and solve prevention and emergency control problems.

I do not have the pretentiousness of creating a safety science. However, I propose a set of coherent concepts to help to improve the safety of people, the environment, and properties. People include the readers, their families, friends, coworkers, and community members; the environment includes the atmospheric air, soil, lakes, rivers, seas, groundwater, fauna, flora, and the anthropic environment, and properties include the assets of individuals, organizations, and communities.

In this book, I emphasize industrial and occupational safety. However, the concepts, methodologies, and techniques apply to any organization and people at home or leisure. They also apply to safety in public areas, traffic, nature, and even to criminal acts, as long as one considers their peculiarities.

The application of this book is as a guide for safety studies and basic text for courses and development of safety programs. In particular, it will help managers and supervisors who hold the means to improve safety. To understand it, advanced education is not required since in most cases, I use concepts and knowledge understandable by high school students. We resort to industrial processes, but the examples do not require specialized knowledge and experience to be understood. Regarding the concepts, I define them consistently with people's daily lives, considering their reality and practice.

Therefore, anyone can use this book to promote safety, in particular, engineers, doctors, and occupational safety technicians. Accident prevention committees can use it as a basic tool to guide their expertise. However, safety excellence does not depend only on those professionals but also on managers and supervisors who hold the means to take the necessary actions. However, many of them do not involve themselves with safety, as they should. I believe that the cause lies in the isolated and disintegrated approach that some organizations practice to manage their functions. In order to remedy this deficiency, we must integrate safety with the mission, quality, productivity, environmental preservation, and human development.

I also believe that most managers and supervisors would like to do more to promote safety, but they lack knowledge, methodologies, and techniques. Moreover, when they endeavor to educate themselves, they cannot find a source of coherent concepts and effective instruments. This difficulty also affects safety professionals, who find books on legislation, hygiene and industrial toxicology, reliability engineering and psychological approach, but not a book with a comprehensive approach. I also face this difficulty and intend to contribute to rectifying it.

Knowledge currently available is sufficient to generate better results in the safety field. However, more important than knowledge itself are the relationships we can establish between different disciplines and experiences. Relationships are more important than the parties themselves. Concepts, methodologies, and techniques must catalyze relations. Barriers that arise between people of different backgrounds prevent their knowledge and experiences from "reacting" with adequate speed to solve problems. Like chemistry, which uses catalysts to increase the velocity of reactions, we can use catalysts to remove communication barriers. To break down these barriers, analogies play a fundamental role. Dealing with the organizational issues through geometric, trigonometric, mechanical, and magnetic models, I intend to provide conditions for engineers, administrators, anthropologists, and psychologists to generate ideas with positive synergy. For this reason, I use examples from several fields of knowledge, such as mathematics, physics, chemistry, biology, medicine, psychology, and anthropology. However, deliberately, I avoid embarking in the labyrinths of any specialty to preserve the holistic approach.

The main goal is to provide the conditions to think about safety globally.

However, we can only act locally. Therefore, we must resort to some specialty, but the holistic view must guide both the action and the choice of the specialty itself.

As alphabet letters form words and words form sentences, concepts and techniques must combine themselves, generating ideas, analyses, and problem-solving. Consequently, some chapters are short because they contain only the essential, what other chapters do not have or we cannot obtain by composition. In addition, I avoided "swelling" the chapters with details that one can find in the bibliography.

All the chapters are useful, but let us indicate the main applications. Chapters 1 and 10 are basic for any application, and their study must precede the others. The leader, who wants to intervene in the organization to implement safety management and change the culture, integrating safety with quality, productivity, environmental preservation, and human development, should pay special attention to Chapters 1–4, 8, and 9. Engineers, technicians, and supervisors should study Chapters 5–9 in depth for the application of their safety expertise. Managers who intend to implement safety programs and the engineer or technician who will guide the implementation must analyze this book jointly.

Anticipating possibilities of application, let us present the summary of the chapters, providing an overview of this book.

Chapter 1 presents the holistic approach in safety. Man has been losing the battle against accidents, and the models he developed to understand and interact with the world are the fundamental causes. Neither the Middle Age model, based on faith and the subjective point of view, nor the Cartesian–Newtonian model, based on science and the objective point of view, achieved satisfactory results. A holistic model balances rational and intuitive thinking, harmonizes the ancient Chinese philosophy and the modern general theory of systems, balances the human tendencies for survival and for achieving happiness, and approaches the accident as *a multifaceted phenomenon, resulting from complex interactions between physical, biological, psychological, social, and cultural factors*. On approaching organizations, the holistic method focuses the leadership, management system, and organizational culture. Doing so, the holistic approach arises as an alternative of greater probability of success.

Chapter 2 addresses safety in cooperation with the mission, productivity, quality, environmental preservation, and human development. The adherence concept depicts clients and organization relationships. Addressing administration as science and game, it suggests the application of chess concepts and strategies on administrative issues. It analyzes the relationships between leaders and subordinates and their impact on organizational climate through a shocking comparison between subordinates and apes in captivity. A holistic management whose regulatory function comprises leadership, organizational culture, and management system orients resources as a magnetic field orients magnets. The synchronization of leadership, culture, and management makes the organization able to develop and respond to environmental changes.

Chapter 3 presents a holistic risk management whose fundamental principles state that the accident is a multifaceted phenomenon, resulting from complex interactions between physical, biological, psychological, cultural, and social factors, and that the random, undesirable, and remote nature of the accidents is unfavorable to people engagement in prevention and training activities. To maintain risks below tolerable values and as low as reasonably practicable, its policy states the priority of people integrity over property preservation and assigns the accountability for risk control to those responsible for the activities that generate the risks. To implement an effective management, *thinking globally, acting locally*, presents guidelines, objectives, programs, strategies, methodology, organizational systems, monitoring, indicators, audits, and diagnosis, and addresses safety in all areas, activities, and phases of the life cycle of installations, products, and personnel (professional life).

Chapter 4 presents a holistic emergency management system whose fundamental principles state that the speed of propagation of hazardous events is higher than a man's speed to detect, analyze, and make decisions; under emergency circumstances, a man is likely to fail, and equipment acts faster and with greater reliability than a man does. However, when equipment fails, a man must take action, and training is essential since the emergency control organization must act as a mechanical system. To have an organization able to control emergencies, it establishes the policy, guidelines, strategy, methodology, action areas, and plans. From the deployment of the function emergency control, it develops an organizational structure whose action groups execute functions using resources identified from unfolding their basic functions.

Chapter 5 presents the risk management in interventions in areas and systems for visiting, inspecting, performing modifications, or changing process variables. Interventions introduce impactful agents that may be aggressive, affecting people, the environment, and properties. They may also initiate emergencies or inactivate control systems. Therefore, interventions demand a comprehensive risk and impact control, not limited to the *permit to work* for maintenance services. Stating that who has the means (resources) to control impacts and risks must be the controller, and supported by risk assessment, the risk management in interventions covers both verbal authorizations and the licenses issued by public bodies for the installation and operation of industrial plants.

Chapter 6 presents the methodology for analysis and risk control based on harm generation models. Harm results from the trio: aggressive agent, target, and exposure (first model) or from a demand plus the failure of the emergency control (second model). Identifying a hazard (the quality of aggressive agents), the risk (damages and losses expected over time) results from the product of the frequency by the consequence of a hazardous event. Quantitative, semi-quantitative, or qualitative methods assess frequencies, consequences, and risks. This chapter addresses risk control by analogy with process control. Risk is the controlled variable; the set point is the tolerable risk; and the sensors are the analysis of risks and abnormal occurrences. Knowledge, skills, reliability, and procedures are control variables. The controller can be a person, group, or body that elaborates the action plan based on the deviations from the tolerable risk. The action plan acts on aggressive agents, their targets, and exposure.

Chapter 7 presents techniques for risk analysis, defining objects, foci, methods, auxiliary and complementary techniques, and formularies. It contains the preliminary risk analysis or preliminary hazard analysis, hazard and operability studies, failure modes and effects analysis, What-If, Checklist, fault tree analysis, event tree analysis or risks' series, comparative analysis, interaction matrix analysis, cause tree analysis, planned inspection, and abnormal occurrences analysis.

Chapter 8 presents value analysis in safety. Risk analysis emerged from the military. Value analysis is a civilian creation induced by the limitation of materials' availability during World War II. Its fundamental concept is that of function, and the purpose is to increase the value of objects by executing their functions at lower costs. Value analysis and risk analysis have three elements in common: systematic approach, creativity, and teamwork. In addition, this chapter introduces a new concept that integrates value analysis with risk analysis, the aggressive function that identifies what the object does that can cause damage. The concepts and methods used by value analysis and value engineering, now available to safety studies through the aggressive function concept, make safety part of the set of vital functions of an organization together with the mission, productivity, quality, environment, and human development in a holistic functional diagram.

Chapter 9 classifies human failures in omission, mission, strange act, sequential, and temporal according to the occurrence modes. Whatever the mode, failures occur in three types: technical, carelessness, and conscious: technical, resulting from lack of resources; carelessness, from the inability of subconscious and automatic human mechanisms to control man's actions; and conscious, from decision to apply alternative procedures, i.e., inadequate behavior. Therefore, it approaches behavior control through attitude and behavior consequences, mainly those imposed by culture and leadership. In addition, a value analysis demonstrates that the individual adopts the behavior of higher relative value. Whatever the failure type, four promoters can trigger them: primary, secondary, command, and intruder. Finally, it addresses the common cause failures limiting the reliability of redundant components. Two notable situations demand failure analysis: the emergency and the post-emergency.

Chapter 10 presents an integrated set of 33 basic safety concepts, essential to understand the phenomena involved in accidents, performing risk analysis and elaborating and executing risk and emergency control plans.

1 Holistic Approach

The whole is in the parts and the parts are in the whole.[1]

Everything is interdependent.[2]

Think globally, act locally.[3]

1.1 HARMONIZING THE PARADIGMS

Reducing the frequency and the consequences of accidents is a major challenge to human intelligence. The society invests much physical and mental work and large sums of resources in prevention, but accidents continue to occur, challenging all these efforts. Accidents make no distinction between the most anonymous worker and remarkable people such as John Denver, Ayrton Senna, Rocky Marciano, and even Superman.[4]

Man's battle against accidents presents a remarkable aspect. Apparently, he has sufficient resources to avoid them. Scientific and technological progress created sophisticated methods and devices in various fields of human activity, including accident prevention. However, the main objective remains unachievable, and we witness, perplexed and defenseless, loss of life and physical integrity. Even more remarkable, human factors cause the major part of accidents, and the catastrophic ones occur just where modern technologies are in place, such as Chernobyl and Bhopal.

The human factor that we can observe, record, and quantify is the behavior, the actions that man performs interacting with the world. Attitudes, culture, and the behavior consequences themselves promote behavior, the most visible, immediate, and superficial human phenomenon. The interactions create the opportunities, but the main cause of accidents is man's perception of the world.

Man sees reality through paradigms. A paradigm is a standard set of ideas and values. It is a model for description, explanation, and understanding of reality. The current dominant paradigm, built in the Modern Age and known as Cartesian–Newtonian paradigm, is rationalist, mechanistic, and reductionist.

Rationalist: it prioritizes the reason, the goal, in the belief that the scientific method is the only valid approach to knowledge.

Mechanistic: it sees the universe as a mechanical system, emphasizing quantification, prediction, and control.

Reductionist: it emphasizes the parts, the fragmentation, the disciplinary approach, and the specialist who, in the limit of the specialization, would understand *almost everything from almost nothing*.

Rational thinking is linear, focused, and analytical. In the left cerebral hemisphere resides the functions of reasoning and formal logic, linear and analytical.

In the right cerebral hemisphere resides the functions of understanding, intuition, pattern capture, images, melodies, and poetry, i.e., totalities. Intuitive thinking is a synthesizer, not linear. For our mind, modeled in the Cartesian model, it is easy to accept that the dominant paradigm in the Middle Ages, which mixed reason and faith and whose main objective was to understand the reason of things without exerting prediction and control, has not solved the problem of accidents. It is easy to accept the ineffectiveness of philosophy, mysticism, poetry, subjective, and intuitive.

However, how to accept that the paradigm that has brought us spectacular scientific benefits and technological progress has failed to control one of the most pervasive contributing factors to human misery?

In the late nineteenth century, a cultural change movement began, pointing out the Cartesian–Newtonian model's shortcomings and proposing a new, so-called holistic paradigm, holistic vision, or holistic approach. It is not to deny the current paradigm, but enhance it. Criticism concentrates on imbalance, with an excessive emphasis on rationalism, reductionism, and mechanics. The holistic paradigm has its character expressed in the sentence: *The whole is in the parts and the parts are in the whole*. It contains, in admirable harmony, ancient Chinese philosophy and modern general theory of systems.

The ancient Chinese philosophy sees the reality, whose essence they called Tao, manifesting itself through two complementary opposites: yin and yang. Yin is receptive, consolidator, cooperative, intuitive, and integrative. Yang is expansive, aggressive, demanding, rational, and competitive. Nothing is pure yin or pure yang. Good is not yin or yang, but their balance. Evil is not yin or yang, but their imbalance. In light of Chinese thought, the Cartesian–Newtonian model has an imbalance, favoring the yang.

The modern general theory of systems has in its essence the interrelationship and interdependence of all phenomena. A system is an integrated whole, and its properties result from interactions. Therefore, the properties of the parts do not determine the properties of the system. Systems have their own characteristics, nonexistent in the parts, and relations are more important than the parts themselves. The systemic approach is multidisciplinary.

Living systems have a stratified organization. Each subsystem is a whole to its parts and is a part of a greater whole. Thus, cells form tissues, and tissues form organs, which form larger systems. These combine to form men and women who form families, tribes, societies, and nations. Each subsystem is both a whole and a part, and has two opposing but complementary tendencies: the self-affirmative, preserving the individual autonomy to maintain the stratified order, and the integrative, submitting to the requirements of the greater system to make it viable. A healthy system balances dynamically the two tendencies, which make the system flexible and capable of change. Therefore, the survival of an organization depends on its ability to balance self-assertion and integration in relationships with the larger system and the environment.

Emphasizing excessively the parts and disciplinary approach, the Cartesian–Newtonian model promotes loss of the systemic vision. The mechanistic character regards man as a replaceable gear; living systems, organizations, and ecosystems as machines; and cause–effect relations as deterministic. The reductionist view

Holistic Approach

approaches safety and quality as something not inherent in systems and activities; as such, we may add them. A fragmented analysis leads to non-suitable and ineffective actions. The holistic approach balances the cerebral hemispheres and the yin and yang tendencies, obtaining positive synergy from the two approaches.

Let us consider traffic statistics revealing that drivers cause a high percentage of accidents. The reductionist view would propose actions aimed at changing the drivers' behavior. A holistic approach, however, would also consider the lane's state and the signaling on man's behavior in traffic and on their motivation to maintain vehicles (everything is interdependent!).

The subjective, intuitive, philosophical views are also ineffective. Isolated and disintegrated, they are more detrimental to safety than the Cartesian vision. We believe that there lies a major cause of failure. Cartesian vision creates hazardous technological advances (high velocities, temperatures, pressures). However, consciously or unconsciously, man uses subjective instruments such as "to keep the fingers crossed" to control risks. Moreover, as those instruments fail, they blame the bad luck or fate (*Thirteen is an unlucky number; it was a fatality*). In this context, accidents cause tragedies and create the *seer syndrome*, a perplexity and immobility state. People seek a seer to know the next tragedy without realizing that the accidents occur because people build them. "Traveler, there is no path. The path is made by walking."[5]

1.2 HARMONIZING SURVIVAL AND HAPPINESS

Two fundamental goals drive man: survival and happiness. The unbalance of goals produces disastrous effects. Pursuing survival or perpetuation, man tries to accumulate assets and builds stress and degenerative diseases. Food shortages in prehistoric times developed automatic mechanisms aimed at making most energy available and spending it to a minimum. Despite the current food abundance, man keeps following the *law of the least effort*. As a result, he accumulates fat and sodium, and builds future internal accidents.

In pursuit of happiness, he does not follow the same law. On the contrary, he/she pursues goals of being the first, gaining unrestricted freedom, satisfying every need, violating procedures, and taking dangerous shortcuts, but finds transmissible diseases and accidents of competitions, sporting or not ("The prudent see danger and take refuge, but the simple keep going and pay the penalty").[6]

The survivors are the strongest and the strongest are the sages, not the brave or audacious. Sages establish the balance in pursuit of the fundamental objectives. *The sages have stories to tell.*

1.3 APPROACHING ACCIDENTS

Accidents result from complex interactions between systems' elements and the environment. Their multifaceted nature involves physical, biological, psychological, cultural, and social phenomena. Examples of these phenomena are as follows: (a) physical: machine motion, electric current, heat transfer, light emission, noise, and mechanical shock; (b) biological: movement of people, breathing, sweating, digestion, heartbeat, illness, bleeding, and death; (c) psychological: suspicion, aggression, fear, joy, and anger; (d) cultural: greeting, dress, singing, and ritual; (e) social: strike, absences from work, and election.

Physical, biological, psychological, cultural, and social phenomena are interdependent and interrelated. The cause–effect relations are not all deterministic, but also probabilistic and of correlation.

Our main interest is not on physical, biological, cultural, or social phenomena, but on personal, patrimonial, and environmental damages and losses. Therefore, we want to study the phenomena that cause or can cause damages and/or losses. To perform this study, we need the basic concepts, just as physics has force, mass, and speed, and anthropology has ritual, value, belief, and myth.

Analyzing accidents through *safety concepts*, we perceive automobiles as *aggressive agents*, people as *targets*, gutters as *containment systems*, vehicle actions as *impacts*, effects as *damages*, and streets as *action fields* of aggressive agents. Surprised, we ascertain that agents of such *aggressive capacity* have fragile containment systems and that man can penetrate easily into their action fields. Then, we understand why the traffic accidents have the dimensions we know.

The analysis of multifaceted nature phenomena requires more than a single model. Each model is appropriate to approach a particular aspect of the phenomenon. The holistic approach has two complementary strategies: the reductionist to understand the factors in detail and the systemic to understand the interactions between factors to produce accidents. Let us consider an example: a traffic accident.

The reductionist approach examines each factor in isolation, such as vehicle's state, track and signaling, maintenance system and cultural elements. Beliefs such as "seat belt does not reduce the consequences of accidents," and firmly rooted customs such as disobedience to the signal "stop" are cultural factors.

In turn, the systemic approach establishes relationships between factors, such as "people do not use seat belts" (first cultural factor); "belts are dirty and hidden under the seat" (physical factor); and "the habit of not cleaning belts, and putting them under the seat" (cultural factor). Thus, we form a comprehensive view of the processes that cause a traffic accident.

Let us analyze the interdependencies, interrelationships, and cause–effect relationships between the phenomena involved. Vehicles crash at a crossroad. The first notable phenomenon (mechanical shock) is physical. Damages result deterministically from the shock. The impact throws a driver out of the vehicle, and the other crashes against the windshield. Those physical phenomena result probabilistically from the shock. Cultural phenomena precede the physical ones such as driving without the belt and disobeying traffic signs. Damages have a probabilistic relationship with those cultural phenomena. Psychological phenomena may precede the crash such as inattention and personal problems. Biological phenomena, such as color blindness, hypoglycemia, brain bleeding, and low blood pressure, may cause human failure. Several psychological effects stem from the accidents affecting drivers and their families. Authorities' actions and financial impacts are possible social phenomena.

A holistic analysis does not state that the cause of the traffic accident is the lane state, or the signaling, or the belt, or the driver's psychological state. Accidents have no

Holistic Approach

single cause. In organizations and societies, *the accident is a multifaceted phenomenon, resulting from complex interactions between physical, biological, psychological, social, and cultural factors.*

1.4 APPROACHING ORGANIZATIONS

An organization is a living system. As such, it is part of a larger system, develops itself, and aims perpetuation. Moreover, the organization is in an environment that includes air, water, soil, fauna, flora, and the anthropic environment. Therefore, the study of an organization must consider space, time, and its relations with the larger system and the environment.

The organization is a source of impactful agents that cause effects on the environment. The effects result from both inputs and outputs. Input is everything that enters such as people, raw material, energy, and water. Output is everything that goes out such as people, services, goods, energy, image, and liquid, gaseous, and solid effluents. Some outputs are beneficial; others are harmful. Harmful effects arise from the release of foreign matter into the environment, undesirable transformations, and degradation of inputs such as capturing water and discarding contaminated effluents, receiving healthy people and returning patients.

The effects of the impact can be deterministic or probabilistic. Consider sulfur dioxide released into the atmosphere. Burning fuel oil and knowing its sulfur content, we can determine the amount of sulfur dioxide released. In turn, we cannot calculate the acid amount that accidents spill into a river, but only the probability that a certain amount will spill per year. The risk is a random variable, the expected damage and loss over time.

To study an object, we can divide it into parts according to any convenient criterion. Let us approach an organization dividing it into operating and organizational systems.

1.4.1 Addressing Operational Systems

In operational systems, equipment performs functions, pieces of equipment interact with each other, and man and equipment interact. Operational systems contain persons, equipment, installations, inputs, processes, and products. Persons act as equipment. A man has physical, emotional, and mental characteristics; abilities; experience; knowledge; and creativity, but he executes functions that equipment may come to execute, replacing him.

Inputs include materials, information, and energy. Processes may be continuous or discontinuous. In a continuous process, equipment performs functions without human intervention, except during shutdowns and start-ups. The regulatory function of an operating system is a process specification. The project defines most of the process by linking pieces of equipment. Therefore, modifications in the continuous process imply system modification. The regulatory function of a discontinuous operating system is a procedure. The discontinuous process requires frequent human intervention, and the procedure defines his functions.

1.4.2 Addressing Organizational Systems

In organizational systems, persons perform functions interacting with each other. They have features of living systems, adapting and developing such a maintenance team. Organizational systems may have mechanical characteristics derived from repetitive work or special characteristics such as an emergency control organization and a fire brigade.

Leadership, management system, and organizational culture compose the regulatory function of the organizational system.

Leadership is the position of being the leader of a group, but we also say "the leadership" when referring to persons who lead an organization. When the leadership is formal, the power of influence decreases as one descends in the hierarchical line. When the leadership is informal, no correlation exists between the position and the power of influence since it stems from knowledge and interpersonal skills.

Management system comprises an organizational structure, objectives, policies, strategies, programs, standards, and procedures. The organizational culture comprises values, beliefs, fondness, and proceeding ways. A written document must contain the management system, while culture is necessarily unwritten.

We may divide the organization using criteria of space, time, activities, and functional field. Activities can be organizational, outside the organization, contracted, traffic related, or product related. Each subdivision is an action area of the management system.

1.4.3 Addressing Activities

Some organizations perform activities at defined places, such as the warehouse, laboratory, process unit, and administrative buildings. Some activities, such as welding, painting, and cleaning, occur at any place. Contracted activities require a specific approach since contractors' employees come from different cultures. They know their activity-related risks but ignore the risks inherent in the places they will work. Outside-of-organization activities that employees perform at home, leisure moments and public life, and those traffic related require special risk control strategies. Regarding the consequences of the accidents, they are similar to those occurring in the facilities of the organization, causing absence from work and productivity losses.

1.4.4 Addressing the Life Cycles of the Organization, Products, and Personnel

Conceptual design, basic design (architecture), detailed design (engineering), acquisition (procurement), fabrication, installation, and commissioning are project phases. Start-up, operation, and shutdown are the operation phases. Development, production, marketing, and use and disposal in the environment are the product phases. Selection, training, activities, preretirement, and retirement are the professional life phases.

A phase may introduce a risk factor that will cause accidents in a subsequent phase. Some phases may introduce risk factors that will manifest as accidents in the same

Holistic Approach

phase. Each phase requires specific risk analysis techniques and strategies to keep the risk factors under control and prevent accidents or mitigate their consequences through emergency control.

1.4.5 Making Diagnosis

A study aimed at understanding the safety state of an organization provides a safety diagnosis. Abnormal occurrences, accidents, incidents, damages and losses, and environmental impacts provide inputs. To identify what happened in the past, reports of the abnormal occurrences are the primary source of information. However, as records are not always available or their quality and quantity may be deficient, the critical incident technique must identify what the organization forgot to register. The study must address the organizational climate to assess the organization's emotional state.

Comparing the assessed safety state with a standard one, the study detects deviations, an input for the preparation of an intervention action plan. However, as important as, or even more important than, assessing the deviations is to identify their causes. For this, one needs to assess the operational and organizational systems.

1.4.6 Assessing the Operational System

The study must evaluate the operational system's anatomical (structure and shape) and physiological (functions) dimensions, and address the personnel, equipment, installations, processes, energy, material, operating and maintenance procedures, inputs, products, aggressive agents, and risk control systems.

Aggressive agents can be mechanics (gravitational, elastic, and kinetic), electrical, thermal, biological, ergonomic, sound wave (noise), and radiant energy. Containment, isolation, alarm, protection, evacuation, rescue, combat, and recovery compose the risk control system. Planned inspections assess equipment and systems suitability and integrity, personnel qualifications and competencies, aggressive agents' hazardousness, and control systems quality.

1.4.7 Assessing the Organizational System

Management system assessment focuses on the principles, policy, guidelines, objectives, targets, programs, projects, and procedures. Management system elements must be on paper. If it is not on paper or software, it is not an element of the management system. What people practice, but is not on paper, is cultural element. When something is neither on paper nor in people's practices, it is only a manager's dream.

Leadership assessment focuses on leader's values, priorities, directions, postures, and behavior. Nonverbal language, what leaders do, not what they say, and their image are the key indicators. If safety is a value, the leaders invest time on it. Sentences such as "we deal with the matter when necessary" and "we have meetings where safety matters are also dealt with" strongly indicate that safety is a secondary matter. Specific safety meetings and good image indicate leaders' involvement.

Cultural assessment focuses on beliefs, fondness, values, procedures, heroes, and myths. Reports, stories that the personnel tell, behaviors, practices, clothes, greetings, and responses to questionnaires reveal cultural aspects. Beliefs and values are of the utmost importance in safety diagnosis. They form the attitude, a factor that drives behavior. "Accidents are part of the work" is an unfavorable belief. Investigating the values, one must separate what people say from what they actually think and practice. They may be incongruous if they think their responses will affect them. Therefore, a question such as "Do you value human life?" is naïve. The best indication that something is a value is people's reaction (not speaking, but acting) when facing an offense to that thing.

1.5 PERFORMING INTERVENTIONS

To take the organization from the initial state to a desired final one, the leadership must develop and implement an intervention plan. By physical or financial constraints, the actions must be local, but a global vision is required to understand how the parties will interrelate to produce the desired outcome. The fundamental rule is "Think globally, act locally."

The intervention follows a process control model, comparing the current situation with the desired one, detecting deviations, and informing the controller that defines the intervention actions. The elimination of some deviations requires permanent and flexible action plans, the long-term programs.

A comprehensive action plan promotes an organizational movement, guided by a vision of the future, created by the leadership, and shared with all members of the organization. The change pace depends on the organizational culture, leadership, and management system. The intervention must consider that these components are interdependent, interrelate, interact, and constitute an organizational field that command everything that happens in the organization.

Management system is easy to assess and easy to modify. The intervention involves policies, guidelines, organizational instruments, standards and procedures, programs, and audits.

Safety committees are efficient instruments for intervention. The leaders of the organizational units and their immediate ones constitute functional committees. The *number one* of the organization and his or her management body start the process, extending it until the first-line supervisors and their teams. A functional committee approaches safety issues of its area of responsibility. In addition, cross-functional committees, composed by persons of different areas, impart the global thinking to the components so that they can implement local actions through their functional committees.

Programs must address the organization, contractors and outside-the-organization activities, emergencies, traffic, signage–order–cleanliness, and cultural development. The intervention plan must include permanent instruments, such as hazard identification and risk analysis throughout facilities, products and personnel's professional life cycles, risk control in interventions, and safety monitoring, including analysis of abnormal occurrences.

Leadership is easy to assess and difficult to modify. The intervention involves convincing the top leadership, the *number one* of the organization, that his or her behavior is the key factor to improve safety. Setting a good example is an effective factor to influence behavior. If he or she understands and practices, it is a great starting point. When the leader demands the same practices from his or her direct assistants, it is another great step. A program that includes talks and teamwork to engage leaders is a good tool, but only the day-to-day practice assures the effective involvement.

The leaders either show commitment to safety, by investing time, or show that their involvement does not go beyond speeches.

Culture is difficult to assess and modify. Cultural assessment requires experience. The intervention involves values, fondness, beliefs, and ways of proceeding.

Only leaders can modify values. Coherence between words and acts imparts values to the personnel. Therefore, either the leaders act or the top management must replace them. If the top management does not act, then the desired modification is impracticable.

Changing beliefs demands knowledge and information. Statistical data can demolish unfavorable beliefs. Another strong tool is the opinion of authorities. Authorities are persons that the personnel respect, no matter whether they are drivers, doctors, navigators, or sportsman as long as safety is a concern and what they say has support on actions, not only on theories.

Changing fondness is much more difficult than beliefs. Associating safety matters with persons (or personages and symbols) who count on the audience's fondness can make people attached to those matters.

Changing the ways to proceed stems from the same factors that change behavior, but considering the collective one.

NOTES

1. The essence of the holistic paradigm.
2. The essence of the systemic paradigm.
3. An environmental slogan. It expresses the global interconnectedness of the problems and the virtues of decentralization.
4. Actually, we are referring to the actor Christopher Reeve, who became quadriplegic after an accident (falling from a horse) in May 1995.
5. Antonio Machado, a Spanish poet.
6. Proverbs 22:3.

BIBLIOGRAPHY

1. Capra, Fritjof. 1982. *The Turning Point*. USA: Bantam.

2 Safety at Organizations

... A weak king makes weak the strong people[1]

Luís de Camões, Portuguese poet

2.1 SAFETY FUNCTION

The safety function aims at reducing damages and losses caused by aggressive agents. It is a vital complementary function to the mission of the organization. Therefore, one must integrate safety with the management of the other vital functions. Directing efforts to safety without considering productivity, quality, environmental preservation, and human development is a serious conceptual and strategic flaw. Moreover, we do not improve its performance with exhortations like "safety first," but balancing the vital functions through a holistic management system.

> The absence of holistic view makes the safety function compartmentalized and isolated. The consequences are the disappointments and conflicts.

Every organization has four clients or interested persons: consumer, component, community, and sponsor. The organization meets clients' needs through the mission and the vital complementary functions. Therefore, one must treat these functions with equal importance. Interrelated, interdependent, and interacting, they can present positive synergy. Considering one of them as the most important is misleading. Productivity provides wages (components, employees), profits (sponsor, shareholder), social benefits (community), and resources to develop the organization. Quality conquers consumers and, consequently, generates resources. Safety and environmental preservation prevent injury to persons and damage to the environment and properties, and increase productivity. Human development promotes the performance of any function.

2.2 ORGANIZATIONS

An organization is a set of people with a mission. A group is not an organization. The group can form a community, but to constitute an organization, it must have a mission to accomplish. People waiting for a bus do not constitute an organization. However, when they come together to achieve a common objective, such as making a claim, they form an organization.

As a system, an organization has interrelated and interdependent parts that interact themselves and with the environment, receiving stimuli from outside and developing well-defined-purpose transformations. Organizations are also subsystems of larger systems and as such, one must examine their contribution to the objectives of the larger system.

An organization is a living system, an organism. As such, it has two opposite tendencies: integrative, functioning as part of the larger system, and self-assertion, preserving individual autonomy. Self-assertion behavior is demanding, aggressive, competitive, and expansive (yang). Integrative behavior is receptive, cooperative, intuitive, and conscious of the environment (yin).

The fundamental difference between living and mechanical systems is that the former develops and can adapt to the environment. Living system's order results from coordinating activities that do not constrain rigidly the parts, but leave room for variation and flexibility. This flexibility enables living organisms to adapt to new circumstances. Organizations characterized as an organic system have a continuous adjustment to environmental changes. However, some organizations, such as an emergency control organization, having rigid labor division, clear control hierarchy, and line authority concentrated in the leadership, resemble mechanical systems.

An organization has operational and organizational systems.

In organizational systems, people execute the functions. The main organization may comprise several units or organizational systems. A single person may compose an organizational unit. Leadership, management system, and organizational culture compose the regulatory function of the organizational system.

Equipment, facilities, materials, processes, and products compose the operational system. When the operation is discontinuous, man executes interventions as a component, interacting with equipment and facilities.

An organization acquires special features regarding the existence time, acting ways, and the number of people. The assessment and development of an organization must consider those characteristics.

As for the time of existence (life), the organization may be permanent or temporary. Permanent, when one does not expect its extinction. It exists as long as the larger organization or the community exists, such as a fire department, civil defense, and university. Temporary, when one stipulates a life span.

According to the expected life span, it can be of long, medium, short, or very short duration. A long-duration organization has a life span higher than 1 year such as a group created to deploy a new industrial plant. A medium-duration organization has around 1-year life span such as a company's internal commission for accident prevention (considering that changing all components creates a new organization). It is of short duration when the life span is of months, weeks, or days such as a company's group for maintenance during a shutdown. A very-short-duration organization has a life span of some hours such as a group in a meeting.

Regarding the frequency of actuation, the organization can be of continuous action or by peaks. Continuous action, when the activities do not suffer interruption, such as a laboratory, plant operation, and human resources section. By peaks, when it concentrates the activities in a certain period. The frequency may be high, medium, or low. An orchestra and an emergency control organization actuate by peaks. However, they have a notable difference. Orchestra's peaks are certain since one knows when and where they will occur. Emergency control organization's peaks are random since no one knows when or where they will occur. Consideration of this feature in the conception, design, installation, and development is of paramount importance for the effectiveness of the organization.

TABLE 2.1
Classification of Organizations according to the Number of Components

Category	Number	Components' Relationship
Unitary	1	Working alone, as some professionals
Very small	2–10	They know each other
Small	10–100	Each one knows all the others
Medium	100–1,000	Each one knows 20%–100% of the personnel
Large	>1,000	Each one knows less than 20% of the personnel
Very large	>10,000	Each one knows less than 2% of the personnel

Regarding the number of people, the organization can be unitary, very small, medium, large, and very large. Table 2.1 presents a classification for the organizations and the expected relationships between the components.

2.3 PEOPLE NEEDS

Organizations exist to meet people's needs. The main function of an organization is the mission, the reason for its existence. The mission aims at meeting some need. Hence, the survivability principle is as follows:

> An organization survives while satisfies and does not threaten the satisfaction of people's needs.

People are the essential element of an organization. Equipment and materials perform functions, but people characterize the organization, especially as a living system. Therefore, the performance of an organization depends on people's behavior. Let us consider Abraham Maslow's[2] model to understand people's behavior toward organizations.

According to Maslow, the needs explain much of human behavior. A need activated becomes a stimulus to action, a conductor of the individual activities. It determines what will be important to him and changes his behavior. Therefore, the set of needs is the motivation source (the motive for action). The needs exert guiding or channeling functions. The tension felt when the need is present obliges the individual to exercise the activity. Tension can be pleasant or unpleasant. Latent or unmet needs create discomfort, whose response is the behavior. This aims to reduce tension or discomfort. Reducing tension is a consequence of the behavior. A satisfied need is no longer a source of tension or discomfort. Therefore, only unsatisfied needs are relevant motivation sources.

Maslow developed a scheme, suggesting a system of five basic needs, capable of explaining human behavior largely. He arranged the needs in a hierarchical order. When the individual meets a lower level need, he tends to become aware of the next step. Full satisfaction is not necessary. However, the individual needs a minimum degree of satisfaction to feel free to experience the tensions associated with the next higher need.

Basic needs (physical comfort, shelter, clothing, food) are the lowest level of (Maslow's pyramid) hierarchy. At work, they reflect concerns about pleasant conditions, leisure time, and luxurious personal properties, abstaining from physical exertion or discomfort.

Safety needs constitute the second level. The individual wants to ensure means to achieve basic needs' lasting satisfaction. At work, it is the concern for benefits such as health insurance, retirement plan, safe working conditions, and clear and stable performance standards.

When the individual achieves adequate satisfaction of the safety needs, he reduces the concerns about himself, becomes aware of others, and feels a desire of association. This is the third level, the association's needs. The individual needs to make part of a group and has acceptance's needs. At work, those needs reflect on the concern to have friendly colleagues, the opportunity for interchange, and harmonious interpersonal relationships; to be a team member; and to feel belonging to the organizational family. The individuals expect the organization to provide opportunities for them to meet those needs.

Having satisfied the third-step needs, the individual becomes interested in obtaining special status in the group, feeling the ambition and desire to excel and deserve special recognition. These are the ego-status needs, but we prefer to call recognition needs. They impel persons to seek opportunities to demonstrate competence, hoping to reap the social and professional rewards arising therefrom. The individual feels motivated to give the maximum contribution to the organization in return for many of the rewards that recognition can mean.

When the individual meets the recognition needs, he becomes free to achieve the highest hierarchy level, the self-realization. He wants to prove himself, needs to confirm his own ability, and experiences challenging and meaningful work. This way he seeks to use his creativity and acquires the sense of personal development, fulfillment, and satisfaction through what he does.

Frederick Herzberg[3] addressed the same subject and found that the lower level needs (basic, security, and part of the association needs) represent only hygienic support. They ensure minimal conditions for the health but do not necessarily increase motivation for good performance.

Improving employees' performance through motivation is an application of Maslow and Herzberg's theories. Let us use them comprehensively, considering all the clients of the organization. The role that an organization can play in meeting clients' needs determines their behavior toward the organization.

> Under the holistic view the human needs interrelate, interact, are interdependent and integrated, forming a whole that one must address systemically, and motivation is a multifaceted variable resulting from complex interactions between basic, safety, association, recognition and self-realization human needs.

2.4 ADHERENCE

People and organization relationships are complex as they involve needs, motivations, organizational climate, team morale, integration, and culture. Several interrelated and interdependent factors intervene.

Let us introduce a concept, the "adherence," to facilitate the approach and analysis of these relationships and to depict that concept through an imaginary angle that represents the relations between an individual and an organization, or between

Safety at Organizations

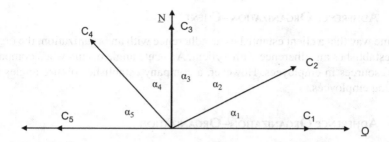

FIGURE 2.1 Adherence client C–organization O.

two organizations. The smaller is the angle, the greater the adherence and cooperation; the greater the angle, the greater is the detachment that can be even an antagonism.

Let C be a client, O be an organization, and α be the adherence angle that C establishes with O. The adherence angle may assume any value between 0° and 180°. Figure 2.1 shows the remarkable cases of adherence between a client C and an organization O and the related adherence angles (α). The projection C/O represents the energy or resources that C places in O; projection C/N (N is the neutral axis) represents the energy or resources that C reserves for other activities.

2.4.1 Adherence Client–Organization

Let us interpret α's values.

α = 0° (C_1). This is an extreme case. It can be detrimental to the client and the organization itself. In it or for it the client puts all his vital energy, such as an employee who gives his life for the company; a businessperson who works day and night and weekends, not paying attention to his family; a community that depends on a single company; and a fanatical football supporter.

α = acute (α < 90°) (C_2). This is the desirable case. Client invests part of his resources in the organization. The projection C/O represents those resources. The complement, projection C/N, he invests in other activities or organizations. An employee dedicates time to leisure and family, the investor invests in more than one company, the community welcomes several companies, and the consumer has several options in the market.

α = 90° (C_3). The client remains neutral toward the organization. He neither puts resources in its favor nor works against it. An individual makes part of a working group without knowing its objectives, or a consumer does not need the product of a company.

α = obtuse (90° < α < 180°) (C_4). The client invests some resources against the organization's interests. An employee that is on strike; a community that suffers effects of company's pollution.

α = 180° (C_5). It is another extreme case. The client puts all his resources against the organization. An army incorporates a soldier who is actually an enemy spy.

2.4.2 Adherence Organization–Client

The same way that a client establishes its adherence with an organization, the organization establishes an adherence with a client. An acute angle occurs when companies invest resources in employees. However, a company establishes obtuse angles with defaulting employees.

2.4.3 Adherence Organization–Organization

An organization also establishes an adherence with another one. Table 2.2 presents some references.

2.4.4 Perceived and Real Adherence

Perceived adherence may be different from the real one. The following cases may occur:

α **perceived** > α **real.** The perceived adherence is lesser than the real one.
α **perceived** < α **real.** The perceived adherence is higher than the real one.
Both cases indicate poor performance in assessing adherence.

2.4.5 Adherence and Quality

Let us analyze the relation between adherence and quality through a geometric figure.

Figure 2.2 represents quality (Q) and client (C) through a geometric circle. If Q increases, α reduces (adherence increases). When Q tends to infinity (in practice, very high quality), α tends to zero meaning total adherence. When Q tends to zero (very low quality), α tends to 90° (indifference). If Q is negative (obtuse angle), C perceives the organization as harmful.

2.4.6 Adherence Control

Knowing the factors that determine the adherence, one can modify it in a planned manner. Let us analyze some significant factors.

Adherence organization–organization. An individual may be client of several organizations. The employee is a component of the sector where he works, and may

TABLE 2.2
Adherence Organization–Organization

Adherence Angle	Typical Case
0°	Supplier having only one customer
Acute	Companies maintaining partnership
90°	Companies operating in different branches
Obtuse	Competitors disputing a restricted market
180°	Two armies at war

Safety at Organizations

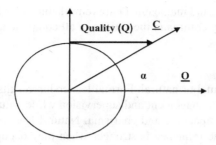

FIGURE 2.2 Adherence and quality.

be a component of company's chorale and/or fire brigade. If that sector establishes an obtuse adherence angle with the chorale, the employee conflicts and must choose one of them. In general, employees adhere to the organization that meets their most urgent needs. He is likely to abandon the chorale fearing low evaluations and no promotions. The same may occur if the individual makes part of the fire brigade.

Therefore, to promote his adherence to the chorale, the top management must promote the adherence of his sector to the chorale (organization × organization). The top manager can call the sector manager and explain him the strategic role of the chorale on company's organization climate.

Passage rites. Passage rite is the ceremony marking people's entry in the organization or community. Every time people pass from an organization to another, what can occur several times a day, they undergo a rite that can be simple or complex. A greeting to the coworkers is a simple ritual. To sing company's anthem is a complex ritual and marks the transition from family organization to company organization.

Incorporation into an organization or community has an initiation rite such as the 15th birthday party or sweet 15 (party), funeral, president of the Republic sworn in ceremony, company's director taking office ceremony, and the incorporation of a new member into the emergency control organization.

The initiation rite exerts a strong influence on adherence. When the organization despises the rites, the individual segregates himself feeling strange in his own organization. Adherence is poor! This occurs when employees start working without an introduction to the other members, and when organizations have no programs aimed at integrating new members.

Organization's poles. Organization's members have several common points, the convergence points or organization poles. Through the poles, they identify other people as companions. Attention, search for information, inspirations, directions, and motivations converge to the poles. Poles are points of concentration of the force lines of the organization.

Let us identify some notable poles. The organization's memory is a pole in the past; the vision is a pole in the future; uniform, symbols, anthem, beliefs, policy, guidelines, venues, functions, and leaders are poles in the present. Some organizational functions are poles of high intensity because their acts affect many components regardless of their needs and concerns. Social and medical services and human resources are notable poles. Persons in those functions have powerful influence on organizational climate. Functions affecting only specific areas have little influence.

Leadership, culture, and management are remarkable poles because, besides promoting adherence, they command the actions of all components.

2.5 LEADERSHIP

Leadership can be formal or natural. Formal leadership results from the position in the hierarchy, such as management and supervision while natural leadership results from interpersonal competence and charisma. Natural leaders are poles, whatever position they hold. The influence is stronger when they occupy a formal position. The position of greater leadership power is that of the *number one*, the organization leader.

2.5.1 LEADERSHIP AND ADMINISTRATIVE GAME

Conducting an administrative process requires science and art. However, the results seem to emerge from a game. The administrator and the chess player have much in common.

"Chess is too much science to be a game and too much game to be science."[4] Anyone with management and chess experience will certainly notice similarities between these two sciences or two games. What Unamuno declared for chess would serve the administration well.

> Management is too much science to be a game and too much game to be a science.

Chess can be a managerial training tool and a management discipline. The most important personage of chess is not the king or the queen, but the chess player. He defines strategies and tactics. In management, leadership plays an analogous role. Therefore, the *number one*, the leader, has a vital and irreplaceable role.

Chess game. Great masters' chess matches have two notable groups. Those that contain violent attacks and spectacular sacrifices, the first group, attract beginners and average players. They happened before the development of the positional game concept. The main objective was attacking the enemy king, employing all the troops, even loosing material, creating weaknesses and delaying his own development. Modern chess matches, the second group, are those of the positional game. At first, they do not cause greater attraction, but analyzing its depths and consequences one can ascertain that they have the same objective, while taking care of the smallest details.

The precursor of positional chess was the American chess player Paul Morphy,[5] who defeated all the masters of his time, including the German Andersen,[6] the greatest player of all times on combinations. Morphy brought new ideas that revolutionized the playing technique. His main concern was to develop his pieces and avoid creating weaknesses before launching attacks. When his opponents began to attack, they found Morphy "casually" with the well-arranged pieces, for defense and quick and decisive counterattacks. He acted for 2 years, and his departure was a major loss to the chess world. The greatest proof of his merit is that Andersen has returned to be world champion. A few years later, another great master, William Steinitz, established the principles of modern chess strategy.

What about the administration? Are we playing as the modern masters or like the romantic ones, launching ourselves to the tasks without planning and risk analysis, trying to gain time (violent attack), creating risk situations (weak points), and suffering accidents (checkmate)?

We believe that the analogies that we will present, together with the methodologies and techniques, will help to analyze the organizational performance and to identify the causes of many hitherto obscure problems and difficulties.

The first movement. The white pieces make the first movement. Theoretically, they should win. Black pieces play to equalize. In management, we can play with the white pieces, making the first movement. We can plan the work and make risk analysis before starting it. Often, however, as in chess, we lose the initiative, starting a job without preparation. The result is productivity, quality, and safety losses. Accidents produce outages, government interventions, and discontent, and degrade the organizational climate. When we realize, we are on the defensive. The opponent (aggressive agent) is ahead.

Mobility. Mobility is the chess theory foundation. One piece is superior to another if oversees more houses, and a position is preferable to another if its pieces can perform a harmonious work and with greater freedom. Placing people in enclosed compartments, sticking and tying them up, is to impair their mobility and reduce their efficiency and effectiveness.

Relative value of the pieces. In chess, the pieces have relative values. Thus, if a pawn is worth one, the knight and the bishop are worth three, castle five, and queen ten. However, the actual value derives from their development. The pieces are worth what they accomplish! Developing them, positioning them, and ensuring their mobility are the general principles to follow. What about the administration? Are people developed, positioned, and mobile, or restricted and poorly positioned? We can find people with knowledge, skills, and experience, but in the wrong positions. No surprise if they contribute little to the organization.

Good bishop and bad bishop. The bad bishop is the bishop forced to defend fixed pawns situated in houses of its color. It loses agility, has limited movements and functions as a pawn. In turn, the good bishop is the one that has freedom and mobility since it runs colored houses opposite of his pawns. The bad bishop is static when it is behind his pawns and dynamic when it is ahead. The good bishop is stronger than the bad bishop is, especially the static.

This concept has administrative applications. First, we need to identify the bishops of the administration. We may establish an analogy with competent people that cannot address the problems because their own team acts as pawns, neither solving nor letting one solving them. In chess, it is better to hand over the pawn; in administration, to dispense whoever is ahead hindering. It is worth remembering that even senior managers may act as pawns.

2.5.2 Relation Leader–Subordinate

The relationships with the leaders have a strong influence on subordinates' adherence. The ideal relationship is a respectful one. Let us designate it as person-to-person relationship, a win–win[7] relationship. However, someone may establish distorted relations that we call person-to-pig and person-to-ape relationships, both win–lose[8] relationships.

Relation person—pig. A few years ago, the newspapers of Campinas-SP, Brazil, reported that in the rural area a pig had killed its owner. Crouched to restore the animal feed, the 60-year-old man could not defend himself. Explanations on the causes of accidents included a popular saying:

> The pig wakes up seven times a night thinking of killing his keeper because he knows that the hand that brings the food one day will bring the machete.

We can apply it to administration. Companies that routinely make dismissals develop in employees the feeling that the pig nourishes against the pig keeper. Wages play the feed role.

Relation person—ape. Captured by man and kept in captivity, the great apes do not have uniform behavior. The gorilla is furious; the orangutan, sad; and the chimpanzee, playful. When the company exerts dominance over employees, develops person × gorilla, person × orangutan, and person × chimpanzee relationships. None is healthy.

2.6 ORGANIZATIONAL CULTURE

A complex of behavioral patterns, beliefs, material, and spiritual values transmitted collectively characterizes every organization. This complex, called organizational culture, consists of social group's expression forms. The way of thinking and living, the uses, customs, beliefs, values, attitudes, rituals, myths, taboos, heroes, histories, art, forms of behavior, habits, and language are part of the culture. These elements are representative of the "world view" or the dominant paradigm of the organization. Organizational culture reflects the way people respond to stimuli. It arises from the need for perpetuation. To achieve this goal, the group adopts a set of premises that they established, discovered, and developed in the learning process, problem-solving, external adaptation, and internal integration. The cultural ballast that identifies it is a basic condition for a social group to feel a community. Otherwise, it would be just a bunch of people, a human conglomerate. Culture is a set of powerful forces that the group exerts on the individual. The organization has its own personality. This stems from its systemic character. The organization is a system and, therefore, has its own characteristics, nonexistent in the parts.

The culture changes by external or internal pressure. The change in thinking involves a new perception. Something that one saw with admiration may come reprehensible, as it has occurred by wearing wild animal's fur coats.

One can induce cultural change. Leadership has instruments to promote change and must lead the process. Their strongest instrument is the example since imitation is a strong learning way, and people tend to imitate leader's behavior. The vision of the future and the management system also change cultural elements, introducing knowledge, information, and practices.

Let us analyze the influence of some cultural elements on safety function.

Beliefs. Belief is something that people accept as true without question. Some beliefs exert negative commands that impair performance of the safety function.

Safety at Organizations 21

People feel demotivated to control risks for not believing in the possibility of interfering with the events. "Nobody dies the day before; suffer accidents is people's fate; accidents are part of the work; bad thing does not happen to me; I have been working this way for so many years and I have never had an accident." The latter has an implicit misguided inductive reasoning: "therefore, it will not happen." It is misleading because the sample is small. Some years or a working lifetime does not constitute a sufficiently large interval since the frequency of major accidents must be less than one accident per thousand years, or lesser.

It is difficult to eliminate deep-rooted beliefs. The intervention to modify beliefs is part of the organizational development. Information, opinion of authorities, and the example of leaders are intervention instruments.

Systematic information on results of risk control undermines the foundation of the old belief, developing a favorable one.

Authority is someone who people consider knowledgeable on a subject. Respect comes from a life dedicated to ideals, great achievements, and charisma such as Jacques Cousteau[9] on environmental issues, and any firefighter or lifeguard on emergency control. People believe in what authority says. The same opinion has no effect if an ordinary person pronounces it, especially if it is a component of the same organization (*No one is a prophet in his country*).

Heroes and myths. The stories that the organization's staff tell reveal who their heroes are. If people that risked their lives, made *great achievements*, and challenged the safety rules are heroes, those who follow the rules are *mere mortals*. Therefore, the organization's history must replace those heroes. The replacement time depends greatly on organization's velocity in renewing its components.

Values. Value is something important for a person, organization, or community. It is difficult to identify the true values because people may lie when they differ from the prevailing social rule. They say a thing, practice another, and create disguises. This process occurred when the European imposed their religion on the slaves. Pretending to accept, the Africans continued to worship their deities.

The performance of the safety function depends on life's value. However, some episodes featured on television networks show that human life is no longer a value.

Leadership exerts a strong influence on organizational values. Let us suppose that in the past of the organization some people have challenged safety standards to achieve production goals. If these people achieved high management positions, the production is the senior management greater value. Culture incorporates this value and written speeches and policies, stating that safety performance has impact on evaluations and promotions are useless.

Fondness. Fondness is a psychic phenomenon that manifests itself as feelings and emotions accompanied by satisfaction, pleasure, and friendship. The fondness arises in person-to-person and person-to-object relations.

Fondness is unfavorable to the safety function. First, accidents are negative and undesirable events, causing people to wander off the subject. Then, safety function requires coercive measures, and applying them without ability creates resentment.

It is easier to change beliefs than fondness. Strategies to make people attached to an object include associating it with celebrities, characters, and people who count on the audience's fondness. Cartoon characters such as Mickey, Minnie,

Woodpecker, and Tom and Jerry have the fondness of most people. Another strategy is to consider that people develop fondness for what they create. Therefore, one can develop the fondness for the object by promoting participation in its creation and development.

2.7 MANAGEMENT SYSTEM

Management is the act of coordinating people's efforts to achieve organization's goals. The management system is a set of interrelated, interacting, and interdependent instruments that the organization uses to plan, operate, and control its activities to achieve goals. Efficient and effective management must assure that people's needs and goals are consistent and complementary to the organization's goals.

Experiment design, value analysis, risk analysis, statistical process control, and problem-solving methods are technical instruments. Principles, goal, strategies, policy, guidelines, organizational systems, programs (projects, objectives, and plans), activities, methods, norms, and procedures are management system instruments.

1. Principles are the basis of the management system. They result from the dominant paradigm's philosophy.
2. Goal is a future state that one wants to attain.
3. Strategy is a way to achieve the goal.
4. Policy is a rule or set of behavioral rules.
5. Guidelines are directions that can restrict possible paths or give general indications. They are more specific than policies and serve to make it explicit.
6. Organizational systems contain relationships between persons because the tasks involve decisions that equipment cannot take. However, considering the development of artificial intelligence, organizational and operational borders became not so clear.
7. Operational systems contain equipment relationships. When persons make part of the system, they work like pieces of equipment. By extension, the system that, although having an intense network of personal relations, presents repetitive and mechanical characteristics of work, is operational. Its tasks are more repetitive than in actual organizational systems.
8. Program is a set of actions developed within a certain action field. It promotes the evolution of the organizations toward the goals. It consists of specific objectives, guidelines, strategies, projects, activities, and action plans.
9. Objective is an intermediate point in the trajectory that leads to the goal.
10. Project is the smallest unit of action or activity that one can plan and evaluate separately and, administratively, implement. It has non-repetitive work features.
11. Activity is a set of actions with repetitive characteristics, used to achieve and/or maintain goals and objectives.

Safety at Organizations

12. Action plan is a set of integrated actions to reach a certain objective with the indication of whom, when, and where they will take place. It can include projects and implementation of activities.
13. Method is a general way to solve problems.
14. Standard is a set of mandatory rules that discipline an activity.
15. Rule is a restriction imposed on procedures, processes, operations, or equipment.
16. Procedure is the detailed description of a process that takes place in batches. It can be organizational or operational.

Each organization chooses a management system from those available or creates their own. Management by objectives, total quality management, and management by guidelines are quite widespread. Let us present the basic elements of a holistic management system.

2.8 HOLISTIC MANAGEMENT SYSTEM

The integration of the organization efforts depends on efficient communication, which requires sharing the holistic conception and a common conceptual framework.

2.8.1 Basic Principles

1. Man has two fundamental goals: survival and happiness.
2. In pursuit of the fundamental goals, man seeks to satisfy five levels of interrelated needs: basic, safety, association, recognition, and self-realization.
3. Organization's survival depends on the satisfaction of four clients: consumer, sponsor (shareholder), components (employee), and community.
4. The value of the organization derives from the performance of six vital functions: mission, productivity, quality, safety, environment, and human development.
5. To achieve an objective, the organization must maintain it and receive client's recognition for the achievement.
6. The survival of an organization depends on its ability to balance two opposing complementary tendencies: self-assertion and integration. The balance is not a point, but a range, and the best position depends on circumstantial factors.
7. The regulatory function of an organization has three components: management system, organizational culture, and leadership.
8. A healthy organization has no great lags between management system, organizational culture, and leadership. Small adherence angles remain, meaning that a component is "dragging" the other and producing development.
9. No client is more important than the others are.

2.8.2 Policy

The five vital functions that accompany the mission must have the same importance level.

2.8.3 Responsibility

The person responsible for the mission is also accountable for productivity, quality, safety, environmental preservation, and human development.

2.8.4 Scope

Productivity, quality, safety, environmental preservation, and human development must be inherent to all activities and life cycle of the installations, products, and personnel (professional life).

2.8.5 Management Methodology

The management of each vital function must be consistent with the holistic management system. Thus, we can establish the productivity management system (PMS), quality management system (QMS), safety management system (SMS), environmental management system (EMS), and human development management system (HDMS). The safety management splits into risk management and emergency management.

A holistic management requires two complementary approaches. The functional approach optimizes the performance of sectoral functions, the subsystems of the organization. The cross-functional approach optimizes the performance of the higher level functions. The functional approach is reductionist, and the cross-functional one is systemic. Formal structure's sectors adopt the functional approach, and the inter-functional committees adopt the cross-functional approach.

The basic management method consists of planning, executing, and controlling. It applies to both the functional and the cross-functional approaches. Planning produces the action plan. The execution puts it into practice. Control consists of measuring, comparing, deciding, and correcting. Measurement is to determine the values of the variables. The comparison requires standards. A decision is a choice between alternatives, which receives the influence of several factors such as the control model, policy, guidelines, resources, and scenery. Correction is the intervention to eliminate the detected gaps.

2.8.6 Programs' Structure

Programs can be sectoral or systemic. The systemic ones focus on vital functions, and their development requires multidisciplinary and multifunctional teams. Systemic programs may address any vital function such as cultural development, productivity, quality, safety, environmental preservation, and human development programs.

2.9 ORGANIZATIONAL CLIMATE

Emotion or dominant emotion at any time or period characterizes the emotional state of a person. Organizational climate is the emotional state of the organization. An emotional state characterizes an organization when it is common to a high number of people, prevailing over any individual emotional state.

Safety at Organizations

The basic emotions are pleasure, sadness, anger, and fear. The intensity of emotions varies between extremes. Thus, pleasure varies from satisfaction to ecstasy, including love and joy; sadness, from disappointment to despair; fear, from shyness to horror; and anger, from discontent to hatred. Dominant emotional states define different climates. Thus, we have a sadness, fear, anger, or pleasure climate.

The organizational climate acts as the sound on guitar strings. Each string has a natural vibration frequency. A guitar string resonates when the frequency of the incident sound equals its natural frequency. By analogy, people have four strings: pleasure, sadness, anger, and fear. Natural vibration frequency varies from individual to individual. Some individuals are anger sensitive while others are pleasure sensitive. Therefore, not all vibrate equally at the frequency of the organization climate.

Internal and external factors produce the climate, e.g., the vision of the future, external threats, country's economic and social situation, people needs' satisfaction, poles of the organization, leadership, management, and culture.

The term climate is proper for long-term conditions. In the short term, it is better to speak of organizational weather. Cheerful organizations may have periods of sadness, the same way as a dry climate region has some rainy days.

As important as describing the climate is to identify and analyze the causes that produced it. Let us use two methods of identification and analysis. The first is to verify if the factors that influence the adherence could have generated the climate; the second, verifying if something contained in six principles established by the thinkers Hobbes and Locke could explain the climate. Let us select some excerpts from their thoughts:

1. "There is no clearer indication of the equitable distribution of anything than the fact that everyone is content with their share." (Thomas Hobbes)[10]
2. "Men do not take any pleasure in one another's company (but rather a great displeasure) when there is no power to keep everyone in respect." (Thomas Hobbes)
3. "In the nature of man, we find three principal causes of discord: the first, competition, the second, distrust, and the third, glory. The first leads men to attack others for profit, the second for safety, and the third for reputation." (Thomas Hobbes)
4. "Man afflicts others, the more satisfied he is, for in this case he tends to show his knowledge and control the actions of those who rule the state." (Thomas Hobbes)
5. "The usurpation consists in the exercise of a power to which others have a right." (John Locke)[11]
6. "Tyranny is the exercise of power beyond right." (John Locke)
7. "Wherever the law ends, tyranny begins." (John Locke)

2.10 RESOURCES OF THE ORGANIZATIONS

Organizations use resources to perform functions and deliver products. Let us consider 12 resources: time, space, energy, material, equipment and installation, knowledge, information, experience, person, ability, procedures/process, and creativity. Figure 2.3 represents the combination of those resources to perform a function.

Some people criticize the term human resources considering that persons are not a mere resource. Let us employ it with another meaning. Human resources are resources that a person possesses and can apply in the organization. Person, experience, ability, knowledge, and creativity are human resources. Person means person-physical, person-emotional, and person-rational. Experience, ability, knowledge, and creativity are independent resources in the reductionist approach. The holistic approach integrates all human resources since a person is an integrated whole.

Who performs functions, must be a qualified individual. Qualification is the knowledge and the set of attributes that enable the individual to perform a function. Generally, a certification process attests qualification. Someone certifies that the candidate has qualifications and provides a certificate.

Let us analyze the resources of an emergency control organization.

Time. Inelastic resource when the emergency is a cardiac arrest! The time available to save a person is 3 min.

Space. Essential for evacuation.

Energy. Firewater pumps require electrical or thermal energies. Diesel's chemical energy moves engines. Biological energy (muscles) drives the person to handle firefighting equipment.

Material. Water, foam liquid concentrate, and chemical powder are firefighting materials.

Equipment and installation. Fire extinguishers, vehicles, Geiger counter, radio, and telephone are pieces of equipment for emergency control.

Knowledge. Knowledge is the theoretical or practical understanding of a matter. Knowledge comprises science and technology, and is in people's minds and books. Firefighting requires knowledge of fire chemistry, fluid mechanics, and dangerous events such as Bleve[12] and Boilover.[13]

Information. Information is relevant data. The firefighting effectiveness in a fuel oil tank requires information on the volume of stored oil, flash point, and tank diameter. Information is in people's memory, books, and computers. While knowledge remains valid for long periods, years, or centuries, the information can change every minute.

Person. "Person" includes people and their physical, emotional, and rational attributes, such as hearing, vision, muscle strength, weight, temperament, and intelligence. One can address the physical person, emotional person, rational person,

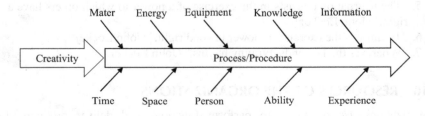

FIGURE 2.3 Resources of the organizations.

knowledge, experience, ability, and creativity separately. We can incorporate skill and experience on this basic human resource. Person-hour and person-day are units to measure the combination of persons and time. Emergency control requires an adequate number of people for each hazardous scenario. Criteria to select a person include height, muscle strength, and other physical, emotional, and rational attributes.

Ability and skills. Abilities are the qualities of being able to do something. Abilities and skills have a small difference. Ability is an innate quality. Skills are the proficiency developed through training or experience. An individual acquires skills learning things such as driving, welding, making decisions, and speaking in public.

Experience. Experience is to know through practice, monitoring, and observation. Brain, documents, drawings, photos, and films store the experience. It is the expertise, what one learns by doing, observing, and testing. It can be individual or organizational. The experience is organizational when widespread and does not disappear when one or some individuals leave the organization.

Knowledge, experience, ability, and skill are different attributes. Consider the welding activity. The engineer studies in books to acquire knowledge on materials' resistance and welding techniques. However, he only acquires experience after months or years, by accompanying welding services, testing, and qualifying welders. Still, it does not develop the skills to weld, unless he performs the welder activity. Moreover, he must have some ability or will never make a high-quality weld.

Procedure. A procedure is a set of instructions detailing how, when, what sequence, and who performs a task. Without written procedures, each individual executes the operation or maintenance his own way. The consequence is the absence of feedback, improvement, and accumulation of experience.

Creativity. Creative includes psychic energy, commitment, dedication, care, willingness to solve, improvement, and pleasure in performing the work. It does not necessarily imply the creation of something new. Such is its importance that we regard it as an independent resource. Persons can have physical, emotional, and rational attributes, knowledge, experience, and skills. However, they will not impart quality, productivity, and safety to their work if they do not put creativity into what they do.

2.11 ORGANIZATIONAL FIELD

The resources of an organization may have four destinations: necessary functions, unnecessary ones, nonuse, and losses. Figure 2.4 indicates the possible destinations for the resources. Necessary functions use the resource R_1 and part of R_2 (projection of vector R_2 on R_1). Part of R_2 is not used, and R_3 has no use at all. Unnecessary functions use part of R_4.

Nonused resources are losses too. Ideally, necessary functions should use all the resources. In the real case, the resources spread in all directions similar to magnets that one pour on a surface. By analogy with the magnetic field, which orients magnets, let us consider an organizational field guiding the resources of the organization. Figure 2.5 depicts the action of the organizational field on resources.

The organizational field is the regulatory function of the organization, resulting from the management system, organizational culture, and leadership. These components interact with one another by modifying themselves.

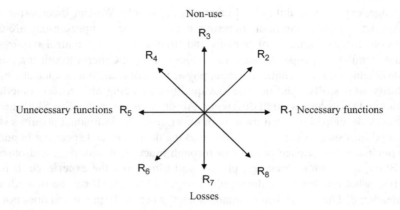

FIGURE 2.4 Resources destination.

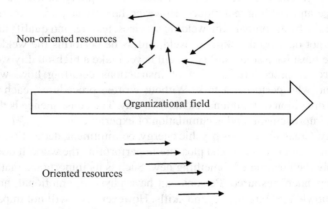

FIGURE 2.5 Organizational field guiding the resources.

The orientation of the resources depends on environmental conditions. By analogy with magnetic reluctance, let us consider that the organizational environment has an organizational reluctance, resultant from several factors, such as organizational climate, complexity of the organizational structure, culture, people's characteristics, and organization poles. The resources have different susceptibilities to the components of the organizational field. Creativity, for example, is susceptible to leadership influence.

Individual and external fields also command the components of the organization. The resulting field is the vectorial sum of the organizational, external, and individual fields. The individual field is internal to each person and may prevail in the command. An external field may also have a strong influence. It results from culture and community laws. Let us represent the resulting field through the following equation:

$$R = MS + OC + L + I + E \qquad (2.1)$$

Where the Organizational field is MS + OC + L = O   (2.2)

where R is the resulting field; O, the organizational field; MS, the management system; OC, organizational culture; L, leadership; I, the individual field; and E, the external field.

Exemplifying a possible conflict between management system and culture, a company may give top priority to pedestrians in its instructions on traffic. However, the culture of the society commands the driver: "Accelerate and advance against pedestrians that cross the street, forcing them to run." Despite the rules, the driver obeys culture's commands.

2.12 ORGANIZATIONAL DEVELOPMENT

Development means evolution. However, evolve to where? First, one must state a goal. The development must promote a change to improve the performance of the organization in a given scenario or to adapt it to new environmental conditions. A vision of the future defines the development trajectory. Let us analyze the trajectory, using the kinematics, the branch mechanics that studies the movements without concerning their causes (forces).

Figure 2.6 depicts the possible trajectories for the development of the organization. O_i is the organization's state, and A_i is its desired state. The organization could reach A_i following the trajectory (1), but it is likely that environmental changes shift the optimum position to A_f, following the trajectory (2). Therefore, the organization must follow the trajectory (3), seeking A, but making route corrections.

The characterization of the state of the organization involves many variables, such as organizational structure, culture, people, policies, programs, expertise, facilities, processes, and products. A notable state is its birth. The French political scientist Alexis de Tocqueville said that the growth of nations presents a common feature. All of them bring some marks of its origin, and the circumstances that accompanied their birth and contributed to their rise affect their entire existence. Let us extend this principle to the organizations stating that we can influence an organization's future taking care of its foundation.

To analyze the speed we can drive the organization's development, let us establish another analogy, with the dynamics, which studies the forces that produce the movements. A dynamics' law states that F = ma, where F is the force, m is the mass, and "a" is the acceleration. Mass measures the body's inertia. The larger the mass, the smaller is the acceleration, for the same force. The mass of the organization

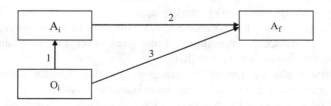

FIGURE 2.6 Organizational development trajectory.

results from dimensional factors such as the number of establishments, persons, and functions.

Another factor limits acceleration by reducing the effective force: friction. The friction force opposes the movement. $F = \mu.P$, where P is the body weight and μ is the friction coefficient that depends on the characteristics of the contact surface. Higher the coefficient, greater the friction force. Organizational friction is the resistance to changes. Culture, climate, and people characteristics determine organizational friction. The friction coefficient of a holistic organization is zero. With a little effort, the organization moves to the desired point.

One can easily shift (develop) a large organization if friction is small. The cult of the small that has spread due to the problems faced by large organizations did not consider that the universe is infinite, and yet it works beautifully. In the Universe, everything interconnects holistically: *The parts are in the whole and the whole is in the parts.* In the real organization, one can accelerate the development reducing the friction forces. The vision of future, policies, guidelines, programs, organizational structure configuration, and cultural development constitute the essence of the organizational universe. These elements make it possible to approximate the real organization of the holistic. Leadership must establish the vision that will guide the organizational development.

Organizational development covers the management system, organizational culture, and the lag between them. In a holistic organization, they are in perfect synchronicity and with total adherence.

Leadership must conduct the organizational development. Management system, by reviewing principles, policies, guidelines, organizational structure, methods, programs, objectives, and goals; organizational culture, by changing values, beliefs, fondness, and ways of proceeding. Leadership itself must develop since what they want to implement may be different from their values, beliefs, and behavior.

Let us analyze the several notable positions of the adherence angles that may occur between the management system and organizational culture. One may use Figure 2.1 and follow the analysis, replacing "O" by the management system and "C" by the organizational culture.

$\alpha = 0°$. The agreement is total. It indicates that the management system and cultural elements have perfect synchronization. On the other hand, it may indicate stagnation, since development requires some lag.

$\alpha =$ acute. Culture is lagging behind the management system. A small lag indicates development. When new guidelines and policies incorporate the management system, people take some time to change the way they behave. Low adherence may occur if these policies and guidelines are different from cultural rules. The lag is significant if the implementation is inadequate.

$\alpha = 90°$. No conflict, but no cooperation. This may occur when organizations operate near complacent communities. Employees never say no when receiving instructions, but also do not execute them.

$\alpha =$ obtuse. Culture reacts to the management system. Conflicts and resistance to changes occur when people perceive the management system as a threat to their needs. It is relatively easy to change policies and guidelines but is not as easy to modify beliefs, values, and rooted customs.

$\alpha = 180°$. This is an extreme case. Culture rejects the management system, generating major conflicts. This may occur when the management system causes significant losses to the satisfaction of people's needs.

Similarly, we can analyze the relationship between leaders and management system, and leaders and organizational culture.

NOTES

1. "... Que um fraco rei faz fraca a forte gente." (Os Lusíadas). Referring Portugal's defeats, caused by their king's weakness, Camões stresses the importance of the leadership.
2. Abraham Harold Maslow (April 1, 1908 to June 8, 1970) was an American psychologist who created Maslow's hierarchy of needs.
3. Frederick Herzberg (1923–2000) was a U.S. clinical psychologist who proposed the motivation–hygiene theory, also known as the two-factor theory of job satisfaction.
4. Miguel de Unamuno y Jugo (September 29, 1864 to December 31, 1936) was a Spanish philosopher.
5. Paul Charles Morphy (June 22, 1837 to July 10, 1884) was an American chess player.
6. Karl Ernst Adolf Anderssen (July 6, 1818 to March 13, 1879) was a German chess master.
7. Negotiation aimed at making the two parties win.
8. Negotiation aimed at making the other party loose.
9. Jacques Cousteau (June 11, 1910 to June 25, 1997) was a French conservationist.
10. Hobbes, Thomas (April 5, 1588–December 4, 1679), English philosopher, one of the founders of modern political philosophy.
11. John Locke (August 29, 1632–October 28, 1704), English philosopher and physician, known as the "father of liberalism."
12. Boiling liquid expanding vapor explosion is the phenomenon characterized by the sudden depressurization of a vessel containing boiling liquid, causing a sudden vaporization of the liquid mass and explosion of the vapor.
13. Boilover occurs after a long period of burning of oils composed of substances with a wide range of boiling, from light to viscous residues. With the burning of the lighter compounds, the residues from the burning surface become denser than the unburned oil and sink, forming a heated layer that moves toward the bottom. When this "hot wave" reaches the water or water emulsion with oil, overheating occurs followed by almost instantaneous boiling of the water, overflowing the tank.

BIBLIOGRAPHY

1. Hobbes, Thomas. 1989. *Hobbes*. São Paulo: Nova Cultural. (Edition in Portuguese)
2. Locke, John. *Locke*. 1989. São Paulo: Nova Cultural. (Edition in Portuguese)
3. Deming, William Edwards. 1990. *Quality: The Management Revolution*. Brazil: Saraiva. (Edition in Portuguese)
4. Williams, Raymond. 1981. *The Sociology of Culture*. USA: The University Chicago Press.

3 Risk Management

Traveler, there is no path. The path is made by walking

Antonio Machado, Spanish poet

3.1 RISK CONTROL

Risk control comprises two auxiliary functions: frequency control and consequence control. Although risk represents the total loss expected over time, and therefore, risk control encompasses the two auxiliary functions, we may use *risk control* referring to frequency and *emergency control* referring to consequences. Risk control aims at keeping the risks below tolerable values. Emergency control acts when the latent risk factors generate abnormal occurrences.

3.2 RISK MANAGEMENT

Risk management is the management of the risk control function, and risk management system is the set of instruments that the organization uses to plan, execute, check, and act to control its risk. Principles, policy, guidelines, objectives, strategies, methods, programs, and organizational systems are management instruments.

Risk control systems may be simple or sophisticated. The same instruments apply to an industrial unit or a single person.

3.3 PRINCIPLES

1. Accident is a multifaceted phenomenon, resulting from complex interactions between physical, biological, psychological, cultural, and social factors.
2. The random, undesirable, and remote nature of the accidents is unfavorable to the engagement of people in prevention and training activities.
3. Accidents occur because the mind engages with the work and forgets the body.[1]
4. Alone, an individual cannot control the risks of his or her activities.
5. All accidents are avoidable.

The fifth principle is valid within certain limits. Outside those limits, man has no power to avoid accidents. This includes a large meteor crashing the Earth. Perhaps one day it will be possible to intercept them. The third and the fourth principles suggest that every activity needs some kind of external control.

3.4 GOALS

The organization's risk management aims at maintaining the risks related to its activities below *tolerable limits* and at a level *as low as reasonably practicable*.

3.5 POLICY

The policy establishes the behavioral rules and reflects the values of the organization. Let us state the following behavioral rules:
1. People's preservation has priority over property's preservation.
2. The responsible for an activity is accountable for the control of its associated risks.

3.6 GUIDELINES

1. Perform risk analysis in all stages of the life cycle of the installations and products.
2. Perform risk analysis of all activities of the organization.

3.7 STRATEGIES

Strategy is a way to achieve an objective or goal. The holistic approach requires a *global thinking*, but *acting locally*. Therefore, we need to define the objectives that can contribute to achieving the goal, such as *involving people with safety subjects*. Let us state a strategy to achieve that objective.

Leader's involvement involves the personnel. A way to show involvement is to invest time such as establishing a weekday to conduct safety meetings. However, that strategy only works if the meetings happen every week, at the same time, *whatever the weather, rain or shine.*

3.8 ORGANIZATIONAL SYSTEMS

Most organizations have a functional structure whose units are organizational systems. According to the policy: *The responsible for an activity is accountable for the control of its associated risks*, each unit shall control its own risks. However, the principle: *Alone, an individual cannot control the risks of his or her activities*, also applies to a group. Therefore, without external support, they will fail in some way. A holistic management can balance the policy and the principle through safety experts assisting the organizational units and implementing functional and inter-functional committees.

The *number one* of the organization and its management body create the first functional committee. The leaders of organizational units and their immediate ones create committees, extending the process to the first-line supervisors and their teams. These committees address safety issues related to their organizational unit.

People from different organizational units form the cross-functional or inter-functional committees. They conduct safety programs affecting all areas, such as risk in transportation and pastime activities, risk analysis, good work practices, and risk control activities. An inter-functional committee provides the global vision to the functional areas where the personnel takes local actions. *Acting locally* takes into account the particularities of the functional areas, and *thinking globally* integrates them.

Risk Management

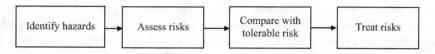

FIGURE 3.1 Risk management process.

In addition, risk management requires some temporary organizational systems. A commission is a short-term organization whose mission is to analyze special problems, such as investigation of an accident. Working group is a short-term organization aimed at performing a technical work.

3.9 METHODOLOGY

The risk management process comprises *identifying hazards*, *assessing risks*, *comparing with tolerable risk*, and *treating risks*. Figure 3.1 presents the diagram of risk management process. Hazard identification and risk assessment constitute the risk analysis. Identification, assessment, and comparison constitute monitoring. Monitoring and intervention constitute the risk control. Control and/or transfer (insurance) constitute the risk management.

The risk management addresses the action areas and the phases of the life cycle of the elements of the organization (personnel, facilities, and products).

3.10 ACTION AREAS

Dividing the organization and activities into action areas facilitates the risk management. The division can be geographic or functional. Thus, we can divide the activities into organizational, outside the work, transportation, contracts, and use of the products of the organization.

Risk management must address off-the-job activities because outside accidents also affect the organization. An employee injured playing football or fishing is an unfit employee. Accidents with family members increase absenteeism. Risks associated with transportation of people and products have characteristics that demand for special approach. Contracts involve cultures and knowledge different from the existent in the organization. Moreover, contract personnel are not familiar with the facilities and their risks, although they must know the risks related to the activities they perform. These characteristics demand a special approach.

We can divide an action area into streets, industrial units, and warehouses (geographic criterion); welding and building maintenance (functional criterion); electricity and compressed air (operational criterion); and phases (life cycle criterion).

3.11 LIFE CYCLE

The conceptual design starts the asset's life cycle; deactivation and disposal into the environment mark its end. Admission starts the employee's labor life cycle; dismissal or retirement marks its end.

During the object's life cycle, the owner must implement risk control measures. Project phases may introduce inherent demands, expose the object to external agents, introduce failure-promoting agents, or failures to install emergency control systems. Control systems aim at maintaining risks below tolerable limits. Every life cycle stage has a suitable risk filter, a system that limits the introduction of risks. It is easier and economical to filter the risks instead of allowing their introduction and implementing additional control measures in the future.

Equipment's life cycle phases are the implementation or project (conceptual design, basic design, detailed engineering, procurement, installation, erection, mechanical completion, and commissioning), operation (production), and deactivation (decommissioning and disposal in the environment). A project is the set of actions aimed at placing an object in the operational phase. Each phase requires specific risk control techniques. Infantile stage, adulthood, and senile are stages of the operational phase. The infantile stage has high failure rates (failures per hour, failures per year) due to failures to assembly, manufacturing defects, or operational inexperience. Free from the infantile stage's problems, the adulthood stage has lower failure rates. However, some new risks arise from wear and tear, overloads, and interventions. Senile stage faces an increase in failure rates due to components' wear. Deactivation can cause environmental damage due to waste disposal.

Product's life cycle phases are the implementation (development, production, storage, transport, and distribution), operation (use), and deactivation (waste disposal).

Employee's professional life cycle phases are the implementation (selection, induction, and training), operation (normal work), and deactivation (preretirement and retirement).

3.12 PROGRAMS

Risk control interventions may have immediate or delayed responses. Changing beliefs and values may require years. Therefore, the process requires long-term action plans, the programs. Specific programs can address organization's activities, activities outside work, transport, contracts, and utilization of products. Due to the multifaceted nature of the accidents, programs require holistic approach through multidisciplinary teams, the cross-functional committees.

In addition to direct action programs, some basic programs support the risk control such as a cultural development program, including *signage–order–cleanliness* and *planned inspections*, addressing all geographical areas, and an audit program, addressing all functional areas of the organization.

Besides these permanent programs, some temporary ones implement and consolidate safety practices such as permit to work, analysis of abnormal occurrences, and risk analysis techniques.

3.13 MONITORING

Monitoring is the systematic verification of attributes of an object through one or a set of parameters. Some parameters are visible, but others require stimulation; some can arise from direct observation or resulting from calculations. Continuous monitoring is an uninterrupted verification. Discontinuous monitoring may be random or at a predefined frequency. Audits, planned inspections, and register and analysis of abnormal occurrences are monitoring tools that require standards (inputs) and detected states and deviations (outputs).

3.14 INDICATORS

An indicator is a qualitative, semiquantitative, or quantitative parameter that represents an attribute. Monitoring produces indicators that can result directly from observations or from formulas, algorithms, or correlations. The risk is an attribute that one can express through an indicator resulting from qualitative, semiquantitative, or quantitative assessment. The risk results from two opposing forces: the hazards and the safety function. Therefore, monitoring activities must address the hazards and the safety function. Hazard indicators include aggressiveness, aggressive capacity, mobility and expansiveness, and exposure and frequency of demand. Safety function indicators include leadership, organizational culture, and management system. The direct indicators of the risk include the number of abnormal occurrences, accidents, damages, and losses.

Risk factors (physical state of the facilities, agents' aggressiveness, behaviors) and abnormal occurrences (accidents, incidents, and near misses) may have deterministic or probabilistic cause–effect relationships or show some correlations. The cause–effect relation is deterministic when an event arises necessarily from a factor; it is probabilistic when an event arises from the factor with a certain probability, and it is a correlation when events and factors have some association, but no detectable cause–effect liaison. Mechanical systems contain deterministic cause–effect relations, and their behavior is predictable. Probabilistic and correlation relations predominate in complex systems such as the ecological and social ones.

Let us present some quality parameters of the monitoring indicators.

An indicator has accuracy or is accurate when is refractory to distortion. Distortion is the difference between indication and reality. Low-severity accidents rate is a low-accuracy indicator when culture promote underreporting. The near miss rates are high-accuracy indicators since people have no reason to hide such occurrences. High-severity accidents rate has high accuracy since it is not practicable hiding such occurrences.

An indicator has sensitivity or is sensitive when detects small variations of the reality. The more the sensitive, the smaller the variations that the indicator can detect. Small variations of the reality cause significant variations in the indicator. Low-severity accidents rate is a high-sensitivity risk indicator while high-severity accidents rate has low sensitivity.

An indicator has a quick response time or is fast when the time to indicate a change of the reality is short. An indicator can be accurate and sensitive, but very slow. A planned inspection is fast to provide risk indicators since it has a low response time while the number of accidents is slow, having a high response time.

3.15 AUDITS

Safety audit is a systematic, documented, and periodic evaluation of the performance of the safety function. Internal teams perform sectoral audits, multi-department teams perform corporate audits, and external teams perform governmental and certification audits.

Generally, audits focus the management system on its elements, i.e., policy, guidelines, programs, action plans, standards, and procedures. They focus too little on leadership, and on culture, even less. The reason is that beliefs, values, and fondness are difficult to assess objectively. Therefore, audits only generate management system indicators.

3.16 DIAGNOSIS

A safety diagnosis results from the verification of the safety aspects of the organization. The diagnosis is an essential input to the preparation of an improvement action plan. While audits focus the safety management, the diagnosis also focuses on hazards, risks, leadership, and organizational culture. Auditing requires a specific protocol, while the diagnosis uses comprehensive safety concepts, standards, and techniques. When the organization implements an improvement action plan based on diagnosis, audits verify that the current actions comply with the plan.

NOTE

1. From a worker of civil construction.

BIBLIOGRAPHY

1. Roland, Harold E. and Moriarty, Brian. 1990. *System Safety Engineering and Management.* USA: Wiley Inter-Science.

4 Emergency Management

Plans are of little importance, but planning is essential.

Winston Churchill

4.1 EMERGENCY CONTROL

State of emergency or emergency is the occurrence of any hazard manifestation. Risk factors emerge from the virtual to the real world, generating damages and losses. Damages and losses result from a chain of events. When the first event occurs, an initiator event or a demand, the emergency starts.

Hazardous events over which man has no control, such as earthquakes and hurricanes, are emergencies. Sabotage acts are emergencies because, despite programmed by someone, they are undesirable events. Operations or activities programmed under controlled conditions for tests and experiments are not emergencies unless something goes wrong.

A top event (explosion, fire, toxic gas leak, or flood) or an initiator event (lack of electricity or cooling water) characterizes an emergency. The *where* (offshore, onshore, city, chemical plant, building, sea) and the *when* (day, night, summer, winter) impart special features to the emergencies as one can ascertain comparing a nighttime with a daytime rescue at sea.

The loss of containment of an aggressive agent is the central element (top event) of an emergency. However, the emergency covers the related causes and effects. An inert gas keeps a reactor's mixture out of the flammable range, but if its control valve fails closing (a demand and an initiator event), the scenario may progress to an explosion.

The risk associated with a hazardous event results from its frequency and consequences. Consequences must dictate the design of emergency control systems. Emergency control includes detection, mobilization, and intervention. Intervention includes the restoration of containment, combat, and defense. An emergency control organization (ECO) is the most complete emergency control system. Planning the implementation and development of such organization requires the identification of credible scenarios, event type, place, moment, and potential damages.

Emergency control function or emergency control is the set of actions aimed at gaining control of situations when risk factors emerge as real facts, threatening to cause damage and loss. Controlling the emergency is to acquire the power to take the situation to a comfortable state. Under control, the release of aggressive agents stabilizes or ceases at all. The emergency control group limits and isolates the aggressive field, preventing additional aggression, rescuing victims, and sending them to medical assistance.

Emergency management system is the set of instruments that the organization uses to plan, operate, and control their activities to control emergencies. Principles, policy, guidelines, objectives, strategies, methodology, programs, and organizational systems are management instruments. The methodology includes identifying emergency scenarios and the initiator event, deploying the emergency control function into auxiliary functions, dividing the organization into action areas, and applying risk analysis techniques, standards, procedures, and monitoring.

4.2 EMERGENCY CONTROL PRINCIPLES

1. The speed of propagation of hazardous events is higher than man's speed to detect, analyze, and make decisions.
2. Under the emergency, man is likely to fail. The probability decreases with adequate training.
3. Equipment acts faster and with greater reliability than man does.
4. It is impracticable to take systems' reliability to 100%. Therefore, when the last mechanical–electronic device fails, emergency control depends on human intervention.

The following guidelines result from those principles:

1. The organization must identify and analyze emergency scenarios and incorporate the critical decisions into action plans so that the individuals can act automatically when demanded to do so.
2. Equipment shall perform as much as practicable the emergency control actions.
3. Personnel who control emergencies must develop skills on fault detection.

4.3 EMERGENCY MANAGEMENT POLICY

Policy is a behavioral rule. It must be simple and objective so that no doubt exist as to the expected behavior. Leadership must establish the policy for the emergency management system. Let us consider the following policy:

> Any hazardous event for people, environment or properties must be the most important event at the time and receive all attention and resources of the organization.

4.4 EMERGENCY MANAGEMENT GUIDELINES

1. Act to protect and without endangering people integrity, including the components of the ECO.
2. Communicate to public agencies (environmental agencies, civil defense) any emergency with potential to affect external areas.
3. Investigate any external abnormality that might be related to the organization's activities.
4. Provide support to the community in emergencies unrelated to the organization activities as long as it does not affect the safety of the organization.

4.5 EMERGENCY MANAGEMENT STRATEGY

The phenomenon emergency is remote, uncertain, and undesirable. These features impart little motivation to prepare for emergencies. The certain, immediate, and positive consequences have strong command over behavior. Therefore, events of minor importance, but with set date and time, have a higher priority than emergency preparedness. What is the solution? A strategy is to create events of certain, immediate, and desirable consequences that prepare the organization for dealing with emergencies. Strategically, the organization invests in attractive events, such as simulated exercises and public presentations, actually aimed at developing skills to deal with real hazardous situations.

4.6 EMERGENCY MANAGEMENT METHODOLOGY

4.6.1 EMERGENCY CONTROL FUNCTION DEPLOYMENT

Equation 6.1 (Section 6.2.1) may guide the deployment of the emergency control function since emergencies are manifestations of risk factors. The equation signalizes that the control must address the agents, targets, and exposure. Figure 4.1 presents the deployment of the main function. Combat group acts on agents while evacuation and rescue prevent exposure, and the medical assistance addresses injured people. Auxiliary functions such as "to provide resources" must meet the needs of the emergency control functions.

4.7 ACTION AREAS

To facilitate the approach, allocation of resources and training, it is convenient to divide the area under the responsibility of the ECO into subareas, such as streets,

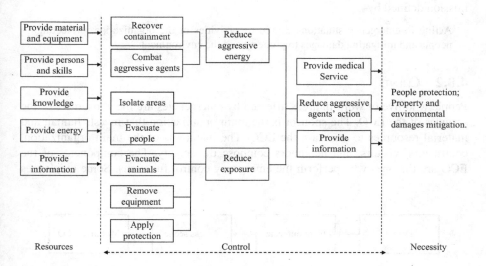

FIGURE 4.1 Emergency control function deployment.

administrative buildings, warehouses, process plants, tankage, laboratory, maintenance workshops, and vegetation.

4.7.1 Emergency Action Plan

Risk analysis identifies hazardous events that demand control measures to prevent evolution to events that are more destructive. Each hazardous event can generate emergency cases when associated with conditions of time and place. Emergency action plan (EAP) must contain scenarios, resources, and procedures. Plans must be simple and straightforward to enable automatic application. Training and exercises must enable the organization to perform in accordance with the plan. The first action is the mobilization. Figure 4.2 shows the process for mobilization of the ECO.

4.8 EMERGENCY CONTROL ORGANIZATION

The ECO performs the emergency control functions. Safety at organizations (Chapter 2) and value analysis in safety (Chapter 8) are the basic instruments used for the development of the organizational model and EAP.

4.8.1 Mission

The identification of hazardous events reveals the need to protect people and mitigate damages to property and the environment. The mission is the basic function of the organization. Let us state the following basic function for any ECO.

> Acting in emergency situations, protecting people and mitigating damages to property and the environment.

In the case of a specific organization, we must delimit the action field, being the mission defined by

> Acting in emergency situations caused by activities of the organization, protecting people and mitigating damages to property and the environment

4.8.2 Clients

People who live or work in areas affected by emergencies are the consumers of the services of the ECO. The sponsor is the main organization that invests human and material resources to maintain the ECO. The integrants of the main organization, contractors, visitors, and neighbors compose the society. The components of the ECO are the ones who perform the emergency control functions. Some employees

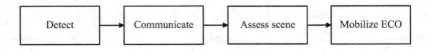

FIGURE 4.2 Mobilization of the ECO.

work exclusively in emergency control (firefighters) and others participate as the components of the action groups (operators, technicians, doctors, nurses, and others).

The success of the ECO stems from clients' adherence. Chapter 2 contains the adherence analysis, a basic instrument to prepare an action plan for development.

4.8.3 Organizational Structure

The ECO must behave as a living organism, developing continuously and acting automatically at critical times. To achieve this goal, we developed a cellular model, using the mechanism of damages generation and the deployment of the emergency control function. Each cell performs its basic function or mission and a set of complementary functions. A cell has a regulatory function consisting of management system, organizational culture and leadership, and 12 resources to perform functions. The nucleus of all cells is equal. The nucleus contains the mission, policy, guidelines, organizational structure, general procedures, and communication system. It connects the cell to the whole.

4.8.4 Resources

The hazardous event is random as to the frequency, location, and moment. It can spread and produce other hazardous events through chain reactions or domino effect. Damages tend to worsen even when they cease the aggressive action and often the hazardous event itself hampers the aid. These features allow us to identify the critical resources of the ECO. In physics, power is the capability to perform work in the unit of time. By analogy, the ECO must be a potent organization, releasing large amounts of actions and resources in the unit of time. Time is an inelastic, scarce, and limited resource that depends on the event itself and its consequences. There is no room to ask for additional time. Skills, materials, and equipment must compensate time shortage. Training is vital. However, agility must be inherent in the organizational model itself.

The resources presented in Figure 2.4 apply to all organization's cells. Combat requires good manipulation of hydrants, foam-launching guns, and firefighting vehicles; public relations, a good internal and external communication; and management, a good decision taking. Products from other cells or external entities meet the cells' needs.

4.8.5 Action Groups

The action groups implement the cells of the ECO. Each organization may adopt its own scheme. Let us present what has proven most effective in our experience. Table 4.1 presents the groups, their basic functions, and some additional clarifications.

The deployment diagram of the basic functions is essential to analyze and prepare the plan for the development of the action groups. The plan involves the allocation of resources, preparation of procedures, training, and drills.

Figures 4.3–4.15 present the deployment diagrams of the basic functions. The captions indicate the name of the group that performs the function in parentheses.

TABLE 4.1
Action Groups' Basic Functions

Action Group	Basic Functions
Management	Take high-impact decisions (Figure 4.3)
Coordination	Coordinate the action groups (Figure 4.4)
Public relations	Provide information to internal and external public. External public includes family, press, visitors, and neighborhood; internal public comprises employees and service providers (Figure 4.5)
Technical	Provide technical advice. It includes firefighting, materials resistance, environment preservation, and legal and insurance issues (Figure 4.6)
Logistic	Provide resources (food, transportation, accommodation, materials, communication equipment, and personnel) (Figure 4.7)
Containment	Contain the aggressive agents. It includes actions such as to stanch leaks of combustible liquids and removing radioactive sources (Figure 4.8)
Medical assistance	Provide emergency medical care to victims (Figure 4.9)
Control	Minimize aggressive actions through combat, removal of victims, evacuation, and isolation. The control group comprises subgroups in Figures 4.9–4.14.
Command	Command the control group (Figure 4.10)
Communications	Provide information. In the ideal case, they have a communications room with telephone switchboard, radio central station, weather station, and database (Figure 4.11)
Brigade	Combat the aggressive agents
	The brigade throws water, chemical foam or powder, halting or reducing heat emissions, toxic gases, and soot (Figure 4.12)
Isolation	Prevent entry into aggressive action field, installing barriers and placing signs (Figure 4.13)
Evacuation	Remove people from the aggressive field, guiding them and accessing a place of safety (Figure 4.14)
Rescue	Rescue victims (people affected or unaffected but in difficulty to leave the aggressive field) (Figure 4.15)

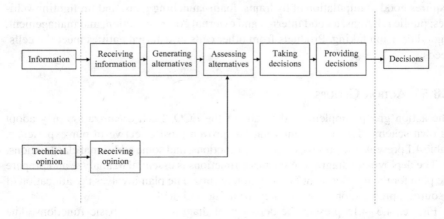

FIGURE 4.3 Functional diagram—Take high-impact decisions (management).

Emergency Management

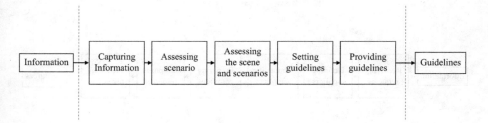

FIGURE 4.4 Functional diagram—Coordinate action groups (coordination).

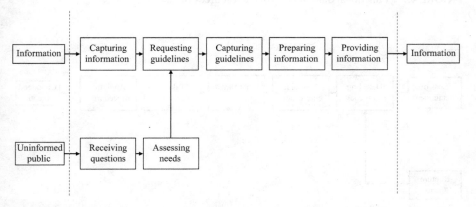

FIGURE 4.5 Functional diagram—Provide information to internal and external public (public relations).

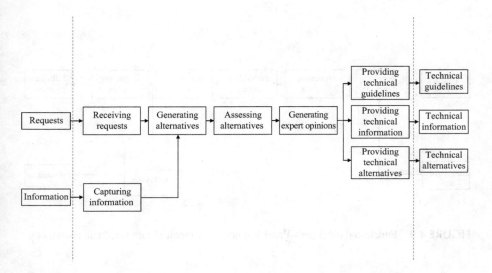

FIGURE 4.6 Functional diagram—Provide technical assistance (technical).

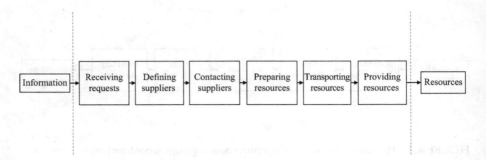

FIGURE 4.7 Functional diagram—Provide resources (logistics).

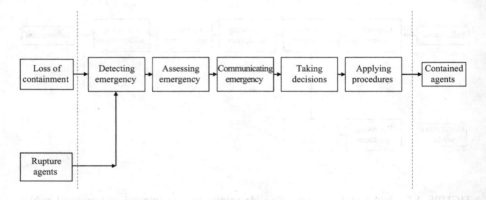

FIGURE 4.8 Functional diagram—Contain the aggressive agents (containment).

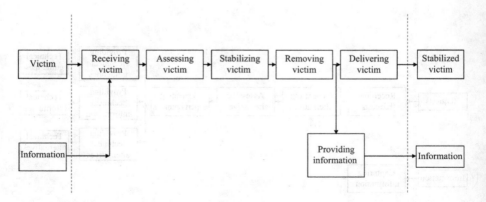

FIGURE 4.9 Functional diagram—Provide emergency medical care (medical assistance).

Emergency Management

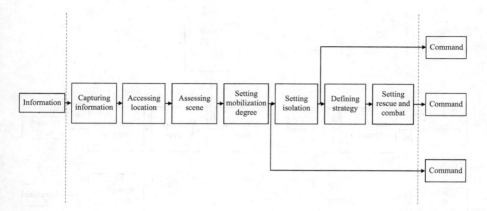

FIGURE 4.10 Functional diagram—Command the combat group (command).

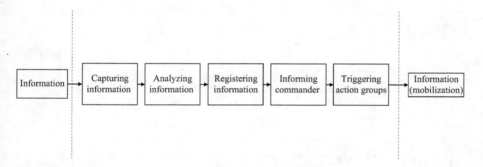

FIGURE 4.11 Functional diagram—Provide information (communication).

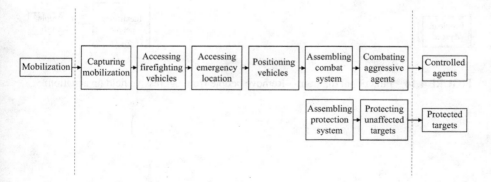

FIGURE 4.12 Functional diagram—Combat aggressive agents (brigades).

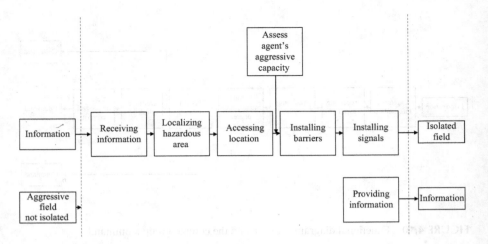

FIGURE 4.13 Functional diagram—Prevent entry into aggressive field (isolation).

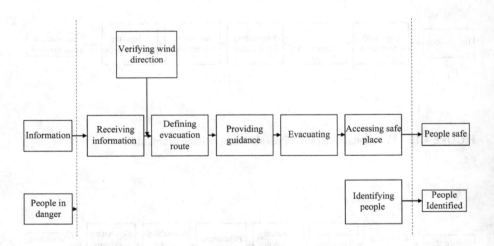

FIGURE 4.14 Functional diagram—Remove people from aggressive field (evacuation).

Emergency Management

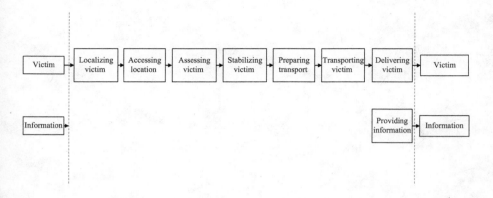

FIGURE 4.15 Functional diagram—Rescue victims (rescue).

BIBLIOGRAPHY

1. Kletz, Trevor A. 2001. *Learning from Accidents*. UK: GPP.

5 Risk Management in Interventions

The military discipline you do not learn, sir, in fantasy, dreaming, imagining or studying, otherwise, seeing, treating and fighting[1]

Luís de Camões, Portuguese poet

5.1 CONCEPT OF INTERVENTION

Intervention is an action directed at an object. Every intervention introduces matter, energy, or information into the object or the environment. Therefore, interventions are invasive, causing impacts and effects. The intervener, object of intervention, environment, impactful agents, targets, impacts, and effects are elements of the intervention.

Aggregates, systems, areas, and processes can be objects of intervention. An area is the region of the space bounded by some criterion. It can be empty or contain aggregates and/or systems. Interventions in systems may affect systemic properties. When the object is a living system, it may be impossible to reconstitute or maintain the systemic properties.

The intervention consists of activities or sets of activities. Activities are sources of impactful agents: persons, animals, equipment, materials, solid wastes, liquid and gaseous effluents, and the various forms of energy.

The impactful agents associated with the intervention act on targets that may be elements of the object, the environment, or the intervener itself. A target can be human, environmental, or patrimonial. The environmental targets comprise the atmospheric air, soil, waters, fauna, flora, and the anthropic environment.

Effects can be immediate or delayed. Regarding quality, the effects can be positive or negative. The maintenance improves equipment's physical conditions and, therefore, the reliability. Harm to the soil and the air, and various types of risk are negative effects.

Interventions may create emergency initiators. Moreover, an intervention can make an emergency control system inoperative.

Interventions may be incidental or unintentional and intentional. This chapter addresses intentional interventions. The unintentional ones have the nature of external demands, and therefore we can address them as careless failures. Due to their random nature, the associated risks are difficult to control.

5.2 TYPES OF INTERVENTION

Every system is a subsystem of a larger system, and every area is a subarea of a larger area. Subsystem's internal activities can affect the larger system or area. The

51

risks generated by interventions from the inside out are difficult to control, because the interveners have restricted knowledge of the larger system or area. Moreover, "It is easier to discard trash through the window." Liquid and gaseous effluents and noise are the products of interventions from the inside out. External interventions are actions coming from outside the system or area, such as visits, inspections, modifications, or maintenance of equipment and facilities, and modification of process conditions.

A visit is someone who enters a certain area to know it. Visitors have limited action and cause minimal interference since someone always guides them. Therefore, the risk created by this kind of intervention is negligible when appropriate procedures are in place.

Inspection is an observation to verify system's attributes. It can be invasive or noninvasive. A noninvasive inspection can be visual or instrumented using techniques such as thermography. Actually, all inspections are invasive. An invasive inspection affects the object of the intervention and in many cases requires partial or total cessation of activities. When the object is a mechanical system, an observation is not an actual intervention. In turn, it is impracticable to observe living systems without intervening.

Modifications, such as building new plants and removing pipes, change the functions of equipment, whether in design or materials, and can affect a system, its components, other systems, and the environment.

Maintenance of equipment and installations, such as an engine repair or replacement of a fuse, aims at recovering normal conditions without changing functions.

The modification of process conditions aims at shutting down, initializing, or changing variables (pressure, temperature, viscosity, composition).

5.3 IMPACT AND RISK CONTROL

The impact control demands the participation of the intervener, who controls the risks inherent in the intervention activities; the *object owner*, who controls the risks inherent in the object; the *area owner*, who controls the risks generated by interactions between elements of the area and the intervention activities; and the *technical advisers* (safety, materials).

Every area, system, or intervention must have a responsible, authority or owner. In process plants, the owner of the area is also the owner of the object of intervention. However, the owner of the interplant area does not own all lines and equipment in that area. Therefore, the risk control during maintenance of pipelines in the area between plants requires the agreement of the area operators, the plant operators, and the maintenance supervisor.

Each responsible person must have the power to control the risks under his or her responsibilities. The power of control results from *authority, responsibility*, and *resources*. Therefore, the control of the risks associated with interventions must consider the following controllability principle:

> If the power of the risk control system is lesser than that of the factors that generate the risks, accidents depend only on random factors.

Risk Management in Interventions

In addition, it must adopt the following strategy:

> Who has the means (resources) to control impacts and risks must act as controller.

Therefore, one must identify the resources that give greater control power. Generally, knowledge and information are the strategic resources.

Inputs, process, products, and regulatory function compose the impact and risk control system.

Inputs are the characterization of the activities of intervention and the description of the object. The characterization must include a description of the area, system, or aggregate and specify what, who, when, and how the organization will intervene.

The risk control process consists of identifying the *hazards*, assessing the associated effects and risks, and defining the measures to control effects, risks, and potential emergencies. To identify hazards and assess the associated risks, one should use formal techniques, such as checklist, preliminary risk analysis, planned inspection, and matrix of interaction. Unprecedented works require open-questioning techniques, while codes, standards, and checklists can control interventions of repetitive characteristics.

The products of the risk control system are the authorization for intervention and the action plan to control the effects and risks. Someone gives an authorization for an intervention at a given time, conditioned to some restrictions. The area owner gives authorization to the system owner, and both give authorization to the intervener. Depending on the expected impacts and risks, the authorization may be verbal or written.

The regulatory function of the organization consists of a standard or organizational procedure. Among other requirements, it must define the cases that require the application of a specific risk analysis technique or the support of specialized technical advice.

5.4 INTERFACES

Areas, systems, components, and environment have common regions, the interfaces. The greater the interface, the greater the difficulty to define responsibilities and greater are the risks. Wherever possible, the interface must be a line. The risks of the interface increase when the resources to control them belong to another area.

Drawings can demark interfaces between areas and systems. Ropes, barriers, painted banners, ribbons, and easels can demarcate interfaces, especially when the borders are temporary. Spades, padlocks, and warning labels can demarcate interfaces between systems (lockout, tagout). Interface risks are greater when one of two systems contain aggressive agents.

By extension, we can define time interfaces. When a project phase ends, the designer transfers the ownership to the constructor who passes to the operator. Transfers must be formal and ritualized so that no doubt remains on who is responsible for the risk control. Otherwise, one transfers the hazards, but not the *resources* to control the risks. Documentation is a key factor to assure that who receives the facilities receives the means to control the risks.

5.5 PERMISSIONS AND LICENSES

Permit to work. Work permission or permit to work is an authorization for intervention that controls maintenance risks. It is an effective practice in accident prevention. However, permit-to-work shortcomings and formalities overvaluing have contributed to catastrophic accidents as the Piper Alpha oil production platform in the North Sea.

When the risk control does not demand a written authorization, it can be verbal, but the principles remain valid. Verbal permission or absence of permission applies when the area owner has no means to control the risks. Electricians have all resources to control the risks associated with replacement of lamps while an office owner has little to contribute. What he can do is stay away from the ladder. Written documents would not improve the risk control. However, the intervention affects office activities and, therefore, requires the authorization of who works there.

Free area. Area owner can relieve the intervener of requesting daily permits when no activities other than the intervention ones are in place, and they do not generate risks beyond the free area. Generally, the issuer imposes some restrictions on specific works that require special risk control measures such as gamma-ray radiography since it affects external areas. Inside the area, the intervener can control the risks of his activities. Area owner gives permission for a certain period, replacing daily permission and declaring the area *free for work* or *free area*. The permit may encompass a piece of equipment instead of the whole area.

Licenses. When the project is an industrial facility, the owner requests governmental authorizations for the installation and operation, such as the license issued by the Environment Protection Agency. Impact and risk control must start at the early project phases (basic design, detail design, procurement). The project may proceed to the next phase if the impacts and risks are below tolerable values. The intervention for installation is the final phase of an industrial project.

NOTE

1. Original in Portuguese: *A disciplina militar prestante não se aprende, senhor, na fantasia; sonhando, imaginando ou estudando; senão vendo, tratando e pelejando.* Experience is a key factor for the performance of the safety function, but in no case, it is as vital as in the control of risks in the interventions.

BIBLIOGRAPHY

1. Kletz, Trevor A. 1999. *What Went Wrong?* USA: GPP.

6 Risk Analysis and Control

You only know the depth of the puddle by falling into it.[1]

6.1 CONCEPT AND METHODOLOGY

Analysis is the division of a whole into parts and the meticulous study of each part. The analyst makes the division according to the criterion that seems convenient. Risk analysis is a detailed study of an object to identify the hazards and assess the associated risks.

Organizations, areas, systems, processes, activities, and interventions can be the object of risk analysis. We can divide areas, systems, processes, operations, and activities, analyzing the resulting subareas, subsystems, functions, and stages. We can divide the overall risk into physical, chemical, biological, and ergonomic risks.

One may use both terms, risk analysis and hazard analysis, since risk analysis comprises hazard identification and risk assessment, and hazard analysis involves risk assessment, even if it is qualitative.

Hazard identification and risk assessment require techniques, such as the preliminary hazard analysis (PHA), hazard and operability studies (Hazop), fault tree analysis (FTA), event tree analysis (ETA), and failure modes and effect analysis (FMEA).

6.2 HARM GENERATION MECHANISM

Let us use two models to explain the harm generation mechanism. The first addresses aggressive agents acting on targets, and the second, the failure of the risk control.

6.2.1 Aggressive Action on Targets

Aggressive agent, target, and exposure are the harm generation factors. Let us consider the following equation:

$$H = (A)*(E)*(T) \qquad (6.1)$$

where H is the harm, A is the aggressive agent, E is the exposure, and T is the target.

The equation says that the aggressive agent produces the harm, but only when a target and exposure exist. No harm is possible when one of them is absent. If one factor is null, the product (harm) is null. Therefore, the control may address one, two, or three factors. Consider a pressure vessel containing ammonia. Harm occurs when ammonia (A) leaks and people (T) are in the aggressive field (E).

6.2.2 FAILURE OF RISK CONTROL

If a demand occurs and the emergency control fails, the hazardous event occurs.

Demands are events requiring emergency control action to avoid the hazardous event. Demand frequency is the number of events in the unit of time (year^{-1}, hour^{-1}). A demand can be inherent in the system, caused by human failure, equipment failure, or external agents.

Demands and failure of control systems are risk factors. To describe the harm generation mechanism, we must answer some questions. Why and how often do demands occur? Why, how, and with what probability do emergency control systems fail?

Persons and equipment create demands when they fail in the strange action mode. A control valve closes despite receiving no command (equipment failure), or when someone improperly closes it (human failure).

Inherent demands do not stem from failures since one cannot eliminate them without affecting the activity itself. A biologist entering into the snakes' pit creates an inherent demand. He prevents the damaging event, snakebite, using personal protective equipment.

The system location creates the opportunity for external agent demands, and one cannot reduce their frequency, as is the case of typhoon, earthquake, and tsunamis.

The emergency control system bars the progress of the hazardous events or reduces their consequences. When demanded to act, the emergency control system may be in a fault state. One can estimate its failure probability as (a) the *fraction dead time* (FDT) (inoperative time fraction) or (b) the historical ratio (number of successful actions)/(number of demands).

Human failure, equipment failure, or absence of control system can cause the failure of a control system. The latter occurs when the project fails to include a control system. When the risk filtration system fails, the project phase introduces a risk factor (failure promoter, absence of control system, or inherent demand).

An organization contains organizational (operations, maintenance, human resources, training) and operational (distillation, storage, chemical treatment, transport) systems. Operational system failures are the immediate causes of accidents; organizational systems' failures are the basic ones. Management system, organizational culture, and leadership constitute the regulatory function of the organizational system.

6.3 HAZARD IDENTIFICATION

Hazard is the quality (property) of everything that can cause harm. Therefore, identifying hazards is to identify hazardous substances, hazardous agents, hazardous products, hazardous situations, hazardous events, hazardous operations, or harmful events. The method and the objectives of the study define the adequate hazard type. However, the risk analysis always demands the identification of hazardous events since one can associate frequencies and consequences to events. Aggressive agents, their sources, targets, release, and exposure possibilities are potential generators of hazardous events. Effective hazard identification requires special techniques, such as Hazop, PHA, and What-If.

6.4 RISK ASSESSMENT

6.4.1 RISK FACTORS

The frequency and consequence of the hazardous event determine the associated risk. Therefore, risk assessment comprises assessing the frequency and the consequences of a hazardous event. Both can be qualitative, quantitative, or semiquantitative. Quantitative analysis requires sophisticated calculation techniques and databases that are neither always available nor reliable. Frequency assessment uses FTA, which requires frequency and probability data of basic events. Consequence assessment requires mathematical models to simulate the phenomena involved. Before embarking on quantitative risk assessment, the analyst must answer some questions:

1. Does the results of the quantitative evaluation justify its costs?
2. Will the control measures stated by the quantitative assessment differ significantly from the qualitative ones?
3. Does the quantitative evaluation differ significantly from the qualitative ones, considering the inherent uncertainties associated with the evaluation of human failures and common cause failures (CCFs)?

Most of the risk control measures do not come from sophisticated calculations, but from a holistic view of safety, considering human failures, CCFs, behavior, *signage–order–cleanliness* (SOC), and *good work practices* (GWPs).

6.4.2 FREQUENCY ASSESSMENT

Before addressing frequency assessment, let us highlight an issue that generates some mistakes, regarding frequency and probability. Frequency is occurrences/time (occurrences/year or year^{-1} or occurrences/hour or hour^{-1}). Probability is a pure number (no units) that assumes values between zero and one. Probability by probability and frequency by probability multiplications are common. However, there is no point in multiplying frequency by frequency, a mistake to which the risk assessment beginners must be aware.

One may assess frequency in two ways. The first is direct, using the historical data. We have two cases to consider.

In the first direct case, we want to know the probability of an undesirable event given that another event, the support event, occurred. For example, the event *engine starts up* creates the unwanted event possibility *engine fails to start up*. The frequency (f_s) of the support event is the number of tentative start-ups in a time interval. The frequency (f_u) of the undesirable event is the number of unsuccessful start-ups in the same time interval. Having experimental or the historical data of f_u and f_s, we estimate the probability (p_u) of the undesirable event through the expression:

$$p_u = f_u / f_s \tag{6.2}$$

Using this probability, one can evaluate the frequency of the undesirable event when the frequency of the support event is available.

When f_s is high, p_u must be very low so that f_u is low. A low probability may be unacceptable if the frequency of the support event is very high since the frequency of the undesirable event may result significant. Let us assume a 10^{-5} probability for the death caused by the yellow fever vaccine, equivalent to one death per 100,000 persons receiving the shot. Therefore, if the health system vaccinates one million persons, ten of them will die, which seems not acceptable for a campaign aimed at protecting people.

In the second direct case, we want to know the frequency of an undesirable event associated with an activity or equipment in continuous operation. This frequency is the number of undesirable events in the time interval (year, hour). Let us consider a washing machine operating for 1 year. We want to know how many failures will occur in 1 year. The support event is the continuous operation of the equipment or the activity. The frequency (f_u) of the undesirable event is the number of times the washing machine stops working in 1 operation year. Having the historical data, we estimate the frequency of the undesirable event as the number of failures in a given time interval. Therefore, if a machine has failed ten times in 4 years, the failure frequency is 0.25 failures/year.

The second way is indirect. Events whose frequencies are as low as once every 100, 1,000, or 10,000 years require indirect quantitative evaluation since the direct evaluation is impracticable. With data from similar devices, we estimate the frequency of failures in failure per device per year or failure per event.

Qualitatively, we can assess frequencies comparing an event with standard events whose frequencies we know from the historical data or opinion of experienced people. Let us set a variable, the frequency level, to facilitate those evaluations.

$$N_f = 10\log(f/f_0) \tag{6.3}$$

where f_0 is the reference frequency assumed as 1.0 occurrence per year.

Table 6.1 presents the frequencies estimated semiquantitatively (occurrences per year) and the respective frequency level for reference events.

Table 6.2 presents the frequencies estimated qualitatively (low, medium, high) for reference events.

Using FTA, we can assess quantitatively the frequency of events that result from basic events whose frequency we know. Major accident hazards studies use this method. Let us analyze an example:

A pressure vessel contains a pressurized gas. A relief valve acts if the internal pressure reaches the maximum allowable pressure. The vessel is the *containment* system, and the relief valve is the *protection* that neutralizes the *rupture agent* (high pressure). The high pressure occurs if two simultaneous events occur, i.e., pressure

TABLE 6.1
Frequency: Semiquantitative Categorization

f (year^{-1})	N$_f$	Reference Events
10^{-8}	−80	Large meteor collides with the Earth
10^{-7}	−70	Liquefied petroleum gas tank ruptures per tank-year
10^{-6}	−60	Fatal accidents by natural agents per person-year
10^{-5}	−50	Highly toxic leaks such as hydrofluoric acid and phosgene
10^{-4}	−40	Large toxic leaks
10^{-3}	−30	Does not occur in facility's lifetime
10^{-2}	−20	Occurs less than once in the facility's lifetime
10^{-1}	−10	Occurs few times in the facility's lifetime
1	0	Occurs at least once a year
10	10	Occurs several times per year
10^2	20	Occurs several times per month

TABLE 6.2
Frequency: Categorization

Category	Frequency	Reference Events
1	Very low	Theoretically possible, but highly unlikely
2	Low	Occurs in special situations
3	Medium	Expected in the activity or installation's lifetime
4	High	Frequent in the activity or installation's lifetime
5	Very high	Very frequent in the activity or installation's lifetime

rises and relief valve fails to open. The first event is a *demand*; the second, a failure of the *emergency control system*; and the third, vessel subjected to a pressure greater than the design pressure, which is the top-level *hazardous event* resulting from the previous ones.

Let us consider a frequency of demand of 0.2/year (one every 5 years) and a frequency of failure of emergency control of 0.01/year (one every 100 years). Valve testing interval is 1 year. Therefore, tests will reveal a failure state once every 100 years on average. As we do not know when the fault state has started, let us assume that it happened in the middle of the test interval, i.e., the valve remains in the fault state for 6 months.

Thus, we have a 6-month fault state in 100 years or an FDT = 0.005. The FDT estimates the probability of the failure on demand. When a demand occurs, the emergency control system may be in a fault state, and the probability is 0.005 or 0.5%. Therefore, the frequency of the top-level hazardous event, vessel under high pressure, is (0.2/year) (0.005) = 0.001 or one every 1,000 years.

When more than one event can cause the top-level event, they have a relation *or* in the fault tree, meaning that each of them is sufficient to cause the hazardous event. Therefore, one must add their frequencies to calculate the frequency of the hazardous event.

Frequency quantitative assessment helps to understand certain terms, such as *in danger* and *at risk*. Let us consider the harmful event *dog attacks a man*. The normal situation is *chain holding the dog inside a yard surrounded by a wall with a closed gate* and *man outside the yard*. The dog attack has a low frequency since this harmful event requires two simultaneous events: *man enters the yard* and *dog releasing*. We assume that once somebody is in the yard, the attack is an event certain. Let p_1 be the probability of occurrence of the first event and p_2 the probability of occurrence of the second one. If those events are independent, the harmful event probability, p_h, is $(p_1)(p_2)$, much smaller than p_1 or p_2 since these numbers are smaller than the unity. However, when the man jumps the wall and enters the yard, the first event occurs; the probability of the harmful event becomes approximately p_2, the probability of dog releasing which is much larger than $(p_1)(p_2)$. Then, the dog attack is much more likely. At that point, the risk reaches a much higher value than the average expected for a longer period. Therefore, the situation is dangerous; the man is in danger or at risk. We should also consider that the two events have some degree of dependency since the dog may break the chain when the man invades his territory.

6.4.3 Consequence Assessment

Consequence analysis evaluates the aggressive agent action field, calculating the aggressive capacity at each point. The study requires mathematical models, and the difficulties to obtain high fidelity results are not few. Analyzing consequences, one must choose the appropriate level of the hazardous event. Fire and explosion are the hazardous events of interest when a flammable liquid leakage occurs. Fires can be pool fire, jet fire, flash fire, and fireball. An explosion may be a deflagration or detonation. Thermal radiation causes an incidence rate, measured in $kcal/h \cdot m^2$, which reduces as the distance from the fire increases. Explosions produce pressure waves whose intensity reduces as the distance from the center increases. If the leaked product is toxic, one must know how it behaves after leaking, as to the direction and concentration at each location. Concentration reduces as the distance from the releasing point increases.

Vulnerability models estimate quantitatively the consequences of the exposure to aggressive fields. The models predict harm to people, environment, and properties through mathematical equations applicable to each event type and affected target. Using the results, one must be careful, especially when addressing toxic substances, since the equations resulted from very limited data or animal experiments.

The severity of the harmful event depends on the capacity of the aggressive agents, harmfulness of the inoculated agent, vulnerability of the target, susceptibility and absorption capacity, and exposure time. Using records of abnormal occurrences to predict the effects, one should work with the expected consequences rather than the reported ones since some events may have caused unlikely damage, different from the expected. For example, the expected harm caused by a bee sting is a swelling and some pain of little gravity. However, people who are allergic to bee stings may die. A risk control system shall consider the average consequence. Otherwise, the control system becomes impracticable since it should prevent bees and people from having any contact. On the other hand, sensitive people should take precautions, and the emergency control system must be able to take prompt actions.

Risk Analysis and Control

TABLE 6.3
Categories of Consequences

Category	Consequence	Characterization	Typical Events
1	Negligible	Insignificant, transient discomfort	Hit an elbow against furniture
2	Very slight	Fast recovery lesions, temporary redness burns	Spraining ankle very slightly, fast contact with hot substances
3	Slight	Slight lesions not causing incapacity to work	Slight twists, cuts caused by sheet of paper, bee stings, contacting hot material
4	Medium low	No permanent damage, recovery in less than 1 week	Light twisting, contacting hot substances, contacting sharp or piercing bodies
5	Medium	No permanent damage, recovery period higher than 1 week	Falls, contacting body or material at high temperature, contacting cutting or perforating devices
6	Medium high	Partial deafness and scars	Exposure to noise of high sound pressure level, body or material at high temperature, action of sharp or perforating bodies
7	High	Blindness, loss of limbs	Hot or very corrosive liquid projection on eyes, sharpening machines action
8	Fatal	One death	Electric discharge, fall from height, inhalation of toxic gases
9	Multi-fatal	Some deaths	Explosions, toxic gas leaks
10	Catastrophic	Large number of deaths	Explosions, large fireballs

Table 6.3 presents a qualitative classification for the consequences of some harmful events. That table applies to humans, but we can construct similar tables for the environment and properties. Specific tables for property damages may be necessary since a catastrophic amount in one case may be negligible in the other.

6.4.4 SEMIQUANTITATIVE AND QUALITATIVE RISK ASSESSMENT

The risk associated with a hazardous event can be assessed qualitatively using Tables 6.2 (frequency), 6.3 (consequences), and 6.4 (risk categories according to frequency and consequence categories). For example, if the consequence category is four and the frequency category is four, the risk category is six. Using Table 6.4, we must consider that subjectivity is inherent in qualitative assessment, and the product of two qualitative parameters is qualitative.

Table 6.5 presents a simpler way to categorize frequencies and consequences, and make a qualitative risk assessment.

The qualitative risk assessment can stem from the quality of the control systems. Tables 6.6 and 6.7 evaluate the quality of the risk control system.

TABLE 6.4
Qualitative Risk Assessment

Frequency Category	Consequence Category									
	1	2	3	4	5	6	7	8	9	10
					Risk Category					
1	1	2	1	1	1	2	2	2	2	2
2	1	2	2	2	3	3	3	4	4	4
3	1	2	3	4	4	5	5	6	6	6
4	2	3	3	6	6	6	7	8	9	9
5	3	4	5	6	7	7	8	9	10	10

TABLE 6.5
Frequency, Consequence, and Risk Category

Consequence	Frequency	Risk Category	
Low	Medium	Very low	1
	High	Low	
Medium	Medium	Medium	3
	High	High	
High	Medium	Very high	5
	High	Extremely high	

TABLE 6.6
Risk Control Instruments

RC[a]	Qualitative	Control Tools
1	Extremely low	Just identify
2	Very low	SOC
3	Low	SOC, GWPs
4	Medium low	SOC, GWP, PIs
5	Medium	SOC, GWP, PIs, procedure, PPE[b]
6	Medium high	SOC, GWP, PIs, procedure, PPE, alarm, isolation
7	High	Project modification
8	Very high	Significant project modification (SPM)
9	Individual extremely high	SPM, questioning any exposure category
10	Social extremely high	SPM, questioning the aggressive agent

[a] Risk Category from Table 6.4.
[b] Personal protective equipment.

TABLE 6.7
Risk Control Instruments: Quality Level

Level	Quality	Characterization
0	Null	Agent's action on targets depends only on random factors
2	Very low	Fails to meet a significant part of standards' critical criteria
4	Low	Fails to meet some standards' critical criteria
6	Medium	Meets standards' critical criteria; fails to meet noncritical ones
8	High	Meets all standards' criteria
10	Very high	Exceeds some benchmark standards

6.4.5 RISK CONTROL INSTRUMENTS

Table 6.6 presents some instruments indicated for each risk category. The quality level of each instrument can be assessed qualitatively using Table 6.7. From these two tables, we can assess qualitatively the risk associated with a hazard and its related hazardous events from the quality level of the set of the indicated risk control instruments.

Another way to qualitatively assess the risk associated with a hazardous event is through the quality of the safety barriers that we can install between an aggressive agent and its targets. Table 6.8 presents safety barriers for risk qualitative assessment with a hypothetical numeric example. Chapter 10 contains the description and concepts related to the safety barriers.

The analysis is specific for each identified aggressive agent. The quality level of any barrier ranges from 0 to 10. If a particular safety barrier does not apply or is not necessary, its quality level is 10. The maximum possible global quality level is 100.

Quality levels 1, 3, 5, 7, and 9 may occur if a safety team does not achieve a consensus about 0 or 2, 2 or 4, and so on.

TABLE 6.8
Risk Control through Safety Barriers

Barrier Number	Safety Barriers	Quality Level (L)
1	Containment	7
2	Restoration	8
3	Isolation	5
4	Alarm	6
5	Protection	7
6	Evacuation	8
7	Rescue	5
8	Medical assistance	5
9	Combat	6
10	Recovery	4
	Global quality level	61

6.5 RISK CONTROL

6.5.1 Process Control

The risk is a variable of the system. Therefore, an overview of process control is useful in understanding risk control.

Controlled or dependent variables are process outputs. Set point is the desired value. In risk control, the risk is the controlled variable, and the tolerable risk is the set point. Manipulated or independent variables are process inputs. They are the available freedom degrees to vary the process and maintain the outputs under control. Training hours, training quality, equipment reliability, planned inspection (PI) frequency, and inspection quality, among others, are the manipulable variables.

Disturbance variables are also inputs, but they are unavailable for manipulation. Disturbances make the risk vary throughout the weeks, days, or hours. During a working day, the willingness to cooperate and people's mood varies. News, rumors, people leaving or returning from vacation, equipment or raw material changes, and many other variables, including the accidents themselves, introduce disturbances that tend to alter the value of the controlled variable (risk). Some disturbances do not produce significant variations, while others may even lead to loss of control.

On-off control actuates when the controlled variable reaches limit values. The control of the liquid level of a vessel opens the outlet pipe valve when the level reaches the upper limit and closes the valve when the level reaches the lower limit. Unfortunately, some organizations use on-off risk control, though inadequate. When severe accidents (upper limit) occur, the controller takes corrective measures. Over time, in the absence of serious accidents (lower limit), it relaxes the control measures, and the risks start growing until a new serious accident occurs.

In cascade control, the first controlled variable establishes the set point of a second controlled variable. The level (primary controlled variable) controller of a tank establishes the output flow (secondary controlled variable) controller's set point. The outlet flow controller manipulates the outlet valve opening (manipulated variable). Similarly, the risk (primary controlled variable) controller establishes the skill level (secondary controlled variable) controller set point that establishes the training hours (manipulated variable).

In the proportional control, the controller action (final control element) is proportional to the error or deviation (the difference between the current value of the controlled variable and the set point). The proportional control does not eliminate the deviation, leaving a residual offset. Similarly, in risk control, the greater is the deviation between the assessed risk and the tolerable risk, the greater is the dimension of the control action. However, the controller must take additional action to eliminate the deviation.

The integral control or readjustment acts while persists a deviation between the value of the variable and the set point. Similarly, in risk control, it corresponds to actions of continuous improvement performed to maintain the risk equal to or less than the tolerable risk.

Derivative control amplifies the controller's action, and the amplification is proportional to the speed of the variation. This control mode fits variables that respond slowly to inputs or disturbances due to system capacitance. Let us analyze the control of the oil temperature at a furnace outlet. If the inlet temperature varies, the outlet temperature will vary with some delay. The effect of the control action also presents some delay. For this reason, the derivative action acts according to the speed of the variation of the outlet temperature. Similarly, the risk controller acts when the number of accidents starts increasing. When this occurs, the causes have gained intensity, but the number of accidents does not increase immediately due to the inertia of the system and the probabilistic nature of the accidents. Therefore, the faster the accident rate increases, the stronger and faster the control action must be.

Feedback control compensates the system when the disturbance (d) effects have occurred. Figure 6.1 shows a feedback controller manipulating the variable (m), based on the error (e) and the difference between the controlled variable (c) and the set point (sp). The disturbance generates the error and the control brings the system back to the desired state or set point. Similarly, the risk controller applies corrective measures when the sensors indicate an increase in risk.

Figure 6.2 shows an anticipatory (feed forward) control correcting the system before the effects of a disturbance (d) manifest themselves. The control system measures a disturbance or some related variable and manipulates a variable (m).

Likewise, the risk controller performs risk analyses before operations or interventions and establishes risk control measures. Analyses performed after the introduction of risks are not anticipatory, but feedback, since the risk is present and may not have manifested itself because of its probabilistic nature.

The actions of the classical control loop and the disturbances (not always measurable) influence the output or controlled variable. Often, the difficulty with

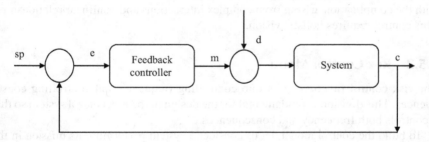

FIGURE 6.1 Feedback control.

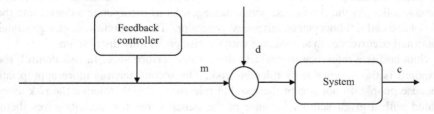

FIGURE 6.2 Anticipatory control (feed forward).

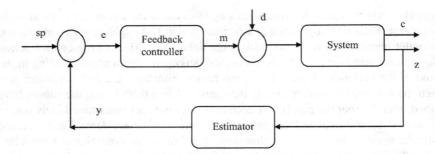

FIGURE 6.3 Inferential control.

this type of control is the measurement of controlled variables and disturbances. This difficulty also occurs in risk control. The inferential control solves these difficulties.

Figure 6.3 shows an inferential control using measurable secondary variables (z) and the manipulated variable (measurable) to estimate the value of the controlled variable through mathematical correlations, obtaining an inference (y). Therefore, it consists of correlating, estimating, predicting, or inferring the value of a variable that one cannot measure. Likewise, the risk controller uses the working environment conditions and unsafe behavior observation to make risk inference and define corrective measures.

Advanced control is a multivariable control that measures the values of several variables and, through calculation algorithms, establishes the changes of the manipulated variables. Analogously, advanced risk control uses many variables, with the complication arising from complex interactions and nonlinear relationships. This control requires holistic vision.

6.5.2 RISK CONTROL MODEL

The risk control function splits into controlling frequency and controlling consequences. This division is fundamental for the design of the risk control system so that it controls both frequency and consequences.

To build the control model, let us consider a system performing its mission in the environment as shown in Figure 6.4. The risk is a product of the system that results from complex interactions between several factors associated with resources, processes, environment, and products. The risk control system contains standards, sensors, and the controller. As the risks cause actual damages, let us incorporate a device into the model and call it "abnormal occurrences' generator." This theoretical device generates abnormal occurrences in accordance with the risk probability distribution.

Standard is a reference to evaluate the system performance. In risk control, the standard is the acceptable or tolerable risk. The second term is more appropriate because people do not accept the risk but tolerate it. People tolerate the risk associated with a given activity because of the benefits that this activity gives them. Employees tolerate better the risks generated by an industry than the neighboring community, as the industrial activity guarantees wages and benefits. The tolerated

Risk Analysis and Control

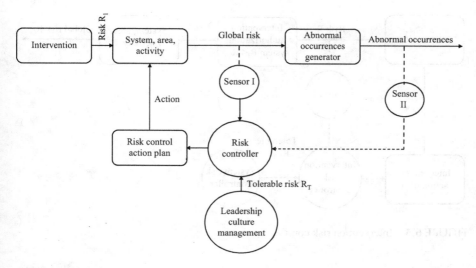

FIGURE 6.4 Risk control.

risk is a management system's key parameter. Social and political scenario coupled with the organization's and the society's economic, financial, and cultural situation have a strong influence on tolerability criteria.

A sensor is a device that measures a process variable. The first sensor evaluates the risk and informs the controller. One must detect the risk variation in the first sensor (type I sensor) so that the controller applies corrective actions, avoiding abnormal occurrences. A second sensor (type II sensor) evaluates the output of the abnormal occurrences' generator. It detects the manifestation of the risks that the first sensor failed to detect. Moreover, it usually detects a large number of occurrences of negligible consequences before a serious event occurs. Waiting for these occurrences to take corrective action is not a good safety strategy.

The type I sensor uses hazard identification and risk analysis techniques, such as PHA, What-If Checklist, Hazop, quantitative risk analysis, PI, FMEA, FTA, comparative analysis, and interaction matrix analysis (simultaneous operations).

Abnormal occurrences analysis reports are type II sensors. A report from similar systems works as a type I sensor.

The controller compares the system performance with standards and introduces corrective actions. To set them, the controller can use one of the control modes. Controller's internal process involves the harm production mechanism, risk assessment, tolerable risk, deviations, leadership orientations, management system, and organizational culture. In practice, the controller is a team or an organ of the organization.

A controller can handle many variables to make the intervention. The choice of the manipulated variable obeys some criterion. Every criterion contains a parameter and a rule. A parameter can be the gain of the variable, and the criterion can be acting on the variable of highest gain. The gain (K) is the variation of the controlled variable per unit of variation of the control variable:

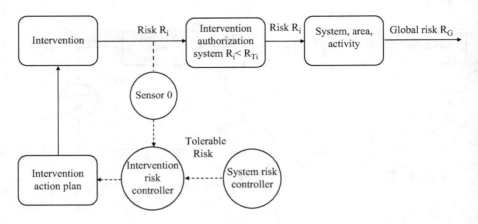

FIGURE 6.5 Intervention risk control.

$$K = \left(\Delta \text{Controlled variable}\right)/\left(\Delta \text{Control variable}\right)^2 \quad (6.4)$$

As we invest in a variable, the gain decreases to new increments. The process achieves the saturation point from which the controlled variable does not change or a further increment has a high cost. Each variable has an optimal point, from which it is better to invest in another. Increasing operators' skills, we reduce the risk. However, when achieving the optimal skill, additional training does not alter the skill or the skill does not alter the risk. Therefore, one should consider other variables such as equipment reliability and the quality of the procedures. Moreover, one must consider that some variables may generate positive synergy, and every change has an associated cost (price). The gain/price ratio is the absolute value of the change, and the controller must opt for the change that produces the greater relative value.

Interventions add new risks to the system, area, or activity. Figure 6.5 shows the intervention risk control system working under the guidance of the general controller.

The type 0 sensor identifies risks before their introduction into the system. They are components of the intervention risk control system.

6.6 ACTION PLAN FOR RISK CONTROL

Risk control action plan or risk control plan contains actions to change the values of the manipulated variables. It is an intervention tool and, depending on the risk size, systems, and organizations involved, it can be simple or quite complex. The plan may contain short-, medium-, and long-term actions.

An action plan for intervention in operating systems addresses immediate causes, while a plan aimed at intervening in organizational systems addresses the root causes.

The models of the harm generation mechanism facilitate the preparation of the action plan.

Risk Analysis and Control

Model I: The harm results from the relation aggressive agent × target.

Let us consider a pressure vessel containing ammonia. The harm results from (a) ammonia leaking, (b) people at the aggressive field, and (c) lack of protection. The control prevents the ammonia leak, removes people from the aggressive field (automation), or eliminates the category IV exposure (protection).

Control on aggressive agent
a. Eliminating the source or reducing the amount and/or the aggressive energy (replacing hazardous substances by inert ones, reducing raw material stocks).
b. Reducing the power of the contributing sources (reducing inventories, flows, pressures).
c. Reducing agents' harmfulness (replacing nonbiodegradable products by biodegradable ones, toxic products by less toxic ones).
d. Reducing leak frequency by increasing reliability (pipes with thicker walls, higher frequency of tests), by installing additional containment systems (storage tanks dikes), containment self-restoration (special valves actuated by the leaking fluid itself), or by combating rupture agents (cathodic protection, relief valves).
e. Combating aggressive agents (diluting of toxic gases, insufflating air, absorbing noise with barriers or filters).
f. Reducing aggressiveness (low voltages, low temperatures) and the harmfulness (eliminating chlorides of nitric acid solutions in stainless steel systems).

Control on targets
a. Reducing susceptibility through selection (people with fair skin should not work in the extraction of marine salt).
b. Reducing vulnerability through selection, design, or construction (blast-resistant control rooms).
c. Increasing the capacity of targets' defense systems (vaccines).

Control on exposure
Reducing exposure probability, time, or category through the following:

a. Appropriate distances so that important targets stay at points where the dilution reduces agent aggressiveness
b. Collective or individual protection systems (acoustic booths, ear protectors)
c. Isolation systems (barriers, boards, standards, training)
d. Sound (siren, beep), visual (plates, colors), and olfactory (flammable gas odorization) alarms
e. Training in alarm detection
f. Reducing access to aggressive action fields (routines, standards, good working practices)

Model II: Damages and losses stem from failures of organizational system.

The controller addresses organizational systems such as a training system, manipulating training hours and training quality.

Management system variables are the policy, guidelines, programs, projects, standards, procedures, and GWPs.

Organizational culture variables are the values, beliefs, fondness, and rituals.

Leadership variables are the posture, commitment, and behavior.

The controller also addresses operational systems. Operational variables related to persons are the physical constitution and number, ability, knowledge, creativity, and experience; equipment and facility variables are the functions, performance, and reliability; process and procedure variables are the relationships between functions, raw materials quality, time, energy, procedure quality, and process variables.

NOTES

1. A popular saying, but we disagree. We can know the puddle's depth without falling into it.
2. Greek letter delta used to express variations.

BIBLIOGRAPHY

1. Lewis, Elmer Eugene. 1987. *Introduction to Reliability Engineer.* USA: John Wiley & Sons.
2. Lees, Frank P. 1996. *Loss Prevention in the Process Industries.* UK: Reed Education and Professional Publishing.
3. Kletz, Trevor A. 1985. *An Engineer's View of Human Error.* UK: Institution of Chemical Engineers.
4. Macdonald, David. 2004. *Practical Hazops, Trips and Alarms.* UK: Newness.
5. Kletz, Trevor A. 1999. *What Went Wrong?* USA: GPP.

7 Risk Analysis Techniques

7.1 PRELIMINARY RISK ANALYSIS

7.1.1 PRA—Object and Focus

Preliminary risk analysis (PRA) is a hazard identification and risk analysis technique that identifies hazardous events, causes, and consequences and states risk control measures. The name preliminary stems from its use for the first approach to study an object. In many cases, it is sufficient to establish effective risk control measures. The object can be area, system, procedure, project, or activity, and the focus are the hazardous events or any undesirable event. PRA and preliminary hazard analysis are names of the same technique.

7.1.2 PRA—Method

   a. Approach the object making the appropriate subdivision.
       If the object is a process, establish the functional diagram and analyze each phase. For example, analyzing a trip, one may consider the travel to the airport, flight, transfer to a hotel, and stay at the hotel.
   b. Select a subdivision.
       Travel to the airport.
   c. Select a hazardous or undesirable event.
       Arrive late at the airport.
   d. Identify the possible causes.
       Taxi breakdown, leave home late, heavy traffic
   e. Identify the consequences.
       Falls, no time for farewells, to miss the flight
   f. Establish risk and emergency control measures.
       Use a checklist to avoid forgetting something, leave home in advance, select a good taxi with radio for quick replacement in case of failure.
   g. Select other hazardous event and repeat the process (back to c).
       Baggage excess
   h. Finishing the hazardous events, select another phase and repeat the process (back to b).

7.1.3 Auxiliary Techniques

   a. Checklists identify aggressive agents and targets
   b. Analysis of abnormal occurrences reports identifies similar hazards that manifested in the past
   c. Planned inspection identifies hazards of an existing facility or activity

7.1.4 COMPLEMENTARY TECHNIQUES

a. Fault tree analysis (FTA) studies the causes with depth
b. Event tree analysis (ETA) studies the consequences
c. Checklist follows the implementation of control measures

The PRA is a good tool to identify hazards and define risk control measures, but it is not practical for direct risk control. In turn, checklists are practical and effective implementation tools.

7.1.5 FORMULARY

The PRA demands a formulary to register hazardous events, causes, consequences, and their category, risk, and emergency control measures. Registering many items can make the form bulky and cumbersome. Therefore, one may list the aggressive agents separately to assist the identification of hazardous events. Table 7.1 presents an example of PRA for a phase of an air travel (going to airport).

7.2 HAZARD AND OPERABILITY STUDIES

7.2.1 HAZOP'S OBJECT AND FOCUS

Hazard and operability studies (Hazop) is a technique for identification of hazards and operability problems.[1] The objects of Hazop are the systems and their processes. Variables such as flow, pressure, temperature, viscosity, and composition define the normal state of a process. Process variable deviations from design intentions are the Hazop's focus. Deviation is the difference between an observed value and the design intention or normal value, such as high flow and low pressure. The set of deviations

---

TABLE 7.1
PRA—Air Travel

Object: Air Travel		Organ:	Page:
Phase: Going to Airport		Number:	Date:
Executed by:			

Hazardous Event	Causes	Consequences	Control Measures
Delays	Taxi breakdown, congested traffic, leaving late	Falls, lack of time for farewells, miss the flight	Take a good taxi, exit in advance, take radio taxi, take cell phone
Car crash	Reckless driver	Injuries	Use seat belt
Baggage excess	Deficient planning	Increasing cost	Define the essential, prepare checklist, weigh the luggage
Forget documents	Deficient planning	Miss the flight	Prepare checklist

contains the subset of hazardous deviations that can act as rupture agents or promoters of aggressive capacity, such as a higher pressure causing the rupture of pipes.

7.2.2 Hazop—Method

Hazop uses guidewords to stimulate creativity. *No, Reverse, Misdirection, More*, and *Less* are the guidewords. Guidewords apply to process variables. *Reverse* applies to variables that can have more than one direction, such as flow and pressure difference. Flow, temperature, pressure, and composition are the most common process variables. The composition includes the expected components and contaminants. Viscosity, density, surface tension, and electric potential appear with lower frequency. The flow can be null (no), going in the opposite direction (*reverse*) or to the atmosphere by leaking (*misdirection*).

Hazop applies to both continuous and discontinuous processes. A continuous process requires the piping and instrumentation diagram (P&ID), and the discontinuous process requires the procedure in the appropriate form. An experienced Hazop leader, a chemical engineer, a mechanical engineer, an instrumentation and control engineer, and an operator constitute the team's core. Additional components join the team according to the process characteristics, but the number should not exceed seven or eight for productivity's sake.

7.2.3 Continuous Process

a. Select a process line.

   Lines and equipment are system's elements. Process line is any link between two main pieces of equipment. Main equipment causes significant changes in the process fluid, such as towers, reactors, and vessels. Pumps, valves, and heat exchangers are line elements. As the choice of the main equipment depends on the criteria of the analyst, the number of lines may be small or large. Many lines make the work tiresome; few lines difficult the hazards identification.

b. Imagine the line operating under normal design conditions. These conditions are references to identify deviations.

c. Select a process variable (flow); apply a guideword to this variable (more).

d. Identify deviations (larger flow) and select the hazardous deviations for analysis.

e. Identify credible causes of the hazardous deviations (valve fails and opens fully).

f. Assess the consequences of the hazardous deviations qualitatively (tank overflows, fire).

g. Check that the operators can detect the hazardous deviation (flow recorder, field indicator, alarm).

h. State the risk and emergency control measures.

   Install a monitoring system for the tank level and implement a training program.

   Install flammable gas detection, firefighting, and an evacuation system.

i. Select another process variable and apply the guidewords.
   Temperature, viscosity, pressure, and composition
j. Identify deviations (higher temperature, lower temperature, higher viscosity, lower viscosity, higher pressure, change in composition, contaminants) and proceed as for the variable flow.
k. Finishing the variables, select another process line and repeat steps (a)–(j).
l. Finishing the lines, select pieces of main equipment and apply the guidewords to the functions they perform and to their process variables. If the function of the equipment is decantation, analyze the deviations: more decantation, less decantation, and reverse decantation (flotation).

Table 7.2 contains an application to a continuous process, the heating system of the feed of a distillation tower.

7.2.4 Discontinuous Process

a. Select an instruction of the procedure.
   Adequate procedures are essential to Hazop effectiveness. The instructions must begin with a verb in the imperative or infinitive mode. They must be short and specific. This way, one can easily apply the guidewords. Procedures are not handouts. A cake recipe should state objectively *to add 100 g of sugar*. We should not make a long explanation why adding sugar or the reasons that lead us to make sweet-flavored cakes.
b. Apply the guidewords to the selected instruction to detect deviations and verify if the identified deviations are hazardous or impair system's operability.

TABLE 7.2
Hazop—Continuous Process

Object: Heating System of Distillation Tower's Feed
Organ: Page:

Executed by: Number: Date:

Variable: Flow

Guideword	Deviation	Causes	Consequences	Control
No	No flow	Human failure closing a valve, control valve fails closed	Overheating, coke formation, tube rupture	Install low-flow alarm, prepare SOP, install fuel cutoff
More	High flow	Control fails, opening valve	Feed pump shuts down resulting no flow	Install fuel cutoff, define testing period for instrumentation

SOP, standard operating procedure.

Risk Analysis Techniques

"More" would lead to detecting the possibility of adding excess sugar. If we add excess sugar, what would be the consequences?
c. Verify that the operators can detect the hazardous deviation.
d. State measures to control risks and emergencies.
e. Select the next instruction and repeat the analysis.

Table 7.3 presents a Hazop for the discontinuous process that caused a major accident in a trichlorophenol industry at Seveso, Italy, September 10, 1976. An exothermic decomposition caused the rupture of a reactor, explosion, and toxic gas emission. The high temperature of the reactor favored the increase of the concentration of TCDD (2,3,7,8-tetrachlorodibenzo-p-dioxin). TCDD is a highly toxic substance, and the accident is one of the most serious occurring in the world.

We analyzed two steps of the procedure used in the reaction system and concluded that a Hazop could have identified the hazards that led to the accident.

The cause of the failure to execute the first instruction could be operator's negligence. Risk control measures could include a checklist to avoid forgetfulness, and the SOP could alert for the identified risks. In addition, the design could provide additional level indicators. Emergency control measures could include a cooling system activated by a high-temperature sensor, evacuation alarm, and alert the community.

7.2.5 Auxiliary Techniques

Hazop requires creativity, a critical resource to detect deviations. Brainstorming techniques promote creativity. However, Hazop does not replace knowledge and experience. Hazop identifies possible deviations, but only those who know the process, the chemical reactions, and have experience can foresee the consequences. Techniques help the memory, but one must have things to remember.

7.2.6 Complementary Techniques

FTA, ETA, and consequences analysis (CA) may complement the Hazop, analyzing in greater depth the identified major hazards and quantify the associated risks.

Checklists control the implementation of Hazop's recommendation.

TABLE 7.3
Hazop—Discontinuous Process

Object: Procedure of the Trichlorophenol Industry

Instruction	Operator Action	Possible Hazop's Find
Distill 50% of the solvent after completing the batch	He distilled only 15%	Guideword: less Deviation: distill less solvent
Add 3,000 L of water to cool the mixture to 50°C–60°C	He did not add a single liter	Guideword: No Deviation: no water added

7.2.7 FORMULARY

Hazop demands a formulary to record deviations, causes, consequences, and risk and emergency control measures.

7.3 FAILURE MODES AND EFFECTS ANALYSIS

7.3.1 FMEA—OBJECT AND FOCUS

Failure modes and effects analysis (FMEA) is a risk analysis technique that identifies failures of components of a hardware system, their modes, effects on the system, the environment, and the component itself. The object of an FMEA is a hardware system and its components. The foci are the failures of the components and their effects.

7.3.2 FMEA—METHOD

a. Select a system (a shower's electrical system).
b. Divide the system into components (wires, circuit breaker, fuse, and resistor).
c. Describe the functions of the components (conduct current, interrupt the circuit, generate heat).
d. Apply the failure modes list to the components, verifying the possible failures (circuit breaker fails to interrupt the circuit).
e. Check the effects on the system, the environment, and the component itself (if the circuit breaker does not open, the fuse burns; but if it fails, the circuit burns).
f. Check the means to detect that the failure is occurring or has occurred (checking the fuse and the circuit breaker).
g. Establish risk and emergency control measures (specify suitable and reliable circuit breaker, maintain spare fuses, and install alert "In case of shutdown do not re-connect all loads").

7.3.3 AUXILIARY TECHNIQUES

a. Functional analysis describes the functions of the system and its components.
b. Failure theory identifies failure modes.
c. Hazop identifies failures by applying guidewords to the functions of the components. Applying the guideword *no* to the function *interrupt electric current* results *failure to interrupt the circuit* (circuit breaker).
d. Checklist monitors the implementation of recommendations.

7.3.4 COMPLEMENTARY TECHNIQUES

a. Event tree analysis identifies the hazardous events that follow the failures (fire).
b. Consequence analysis assesses aggressive fields generated by the hazardous events (escalation to the house).

7.3.5 Formulary

FMEA demands a formulary for registration of the components, their failure modes, detection means, consequences, and risk and emergency control measures. Table 7.4 presents a formulary with the analysis of an electric shower.

7.4 WHAT-IF?

7.4.1 What-If—Object and Focus

What-if is a hazard identification and risk analysis technique that uses open questioning promoted by the question What-if (?). The object can be a system, process, equipment, or event, and the focus is *anything that can go wrong*. The focus is broader than that of other techniques because the questioning is freer. *What-if* is actually a brainstorming.

7.4.2 What-If—Method

What-if admits both free and systematic questioning. Free questioning approaches the object through the question What-if (?) about any hazard or deviation that comes to mind. Therefore, we have questions such as *What-if we add more water (?)*, *What-if we receive contaminated raw material (?)*, and *What-if there is a gale?* The systematic questioning focuses aspects related to disciplines such as electricity, instrumentation, firefighting, environmental preservation, and occupational medicine. In this case, specialists attend to a What-if meeting to answer the questions formulated by the What-if team.

TABLE 7.4
FMEA—Application to an Electric Shower

FMEA—Water Heating by Electric Shower			Organ:	Page:
Object: Shower Electric System			Number:	Date:
Executed by:				
Component	**Failure**	**Effects**	**Detection**	**Control**
Circuit breaker (CB)	Temporal not interrupting circuit in time	Burning fuse	Visual for the fuse	CB quality, resetting procedure
Fuse	Strange action, opening a normal circuit	No current, no heating	Cold water, ammeter, visual	Spare fuses
Cables	Overheating	Power reduction	Visual (color change), odor	Fire retardant cables
Resistor	Rupture	None for circuit, heating fails	Visual for resistor	Resistor quality

7.4.3 Auxiliary Techniques

The systematic questioning demands a checklist.

7.4.4 Complementary Techniques

FTA, ETA, and CA may complement the What-if, analyzing in greater depth the identified major hazards and quantify the associated risks.

Checklists control the implementation of the measures recommended by the What-if.

7.4.5 Formulary

The What-if demands a formulary to register what can go wrong, causes, consequences, and risk and emergency control measures.

Table 7.5 presents the formulary and the analysis of a birthday party.

7.5 CHECKLIST

7.5.1 Checklist—Object and Focus

Checklists verify the conformity of attributes of an object against standards. The object may be area, system, installation, process, and equipment. Lists may have subdivisions by specialty or any other criterion. The focus of the checklist is any deviation from the list. When attributes are functions or their performance, the list contains tests and their standardized answers. An additional use of checklists is to control the risks that other techniques such as PRA, Hazop, and FTA have identified.

TABLE 7.5
What-If (?)—A Birthday Party

What-If?—Hazard Identification

Object: Birthday Party	Organ:	Page:
Executed by:	Number:	Date:
What-If?	**Hazard/Consequences**	**Control Measures**
More people than expected	Lack of space, drink, and food	Consider preparing more food and drink than expected
People do not find the place	Displease friends, dissatisfaction, less gifts	Attach map to invitations and include the phone number
It rains	Difficulties on arrival, people with wet clothes	Acquire large umbrellas to help people move from cars to the house's front door

Risk Analysis Techniques

Checklists are useful and effective when tasks are repetitive since we know the associated risks. Its disadvantage is that the user checks only items included in the list. However, one can enhance its effectiveness using creative techniques whose starting point is on the list.

7.5.2 CHECKLIST—METHOD

The checklist compares attributes of the object with standards.

7.5.3 AUXILIARY TECHNIQUES

a. Functional analysis describes the functions of the object.
b. PRA. Identifies the risks for checklists.
c. Analysis of abnormal occurrences reports provides inputs to prepare checklists.

7.5.4 COMPLEMENTARY TECHNIQUES

Progress control techniques monitor the implementation of corrective measures.

7.5.5 FORMULARY

The checklist demands a formulary containing the items and the checking results.
Table 7.6 presents a checklist for a travel by car.

7.6 FAULT TREE ANALYSIS

7.6.1 FTA—OBJECT AND FOCUS

FTA is a hazard identification and risks analysis technique that can be either qualitative or quantitative. It unfolds a hazardous top event into demands and failures of control systems. The objects of FTA are the systems. Its foci are the hazardous top events and the combinations of events that produce it.

7.6.2 FTA METHOD

FTA is deductive. Starting from a top event as pipe ruptures, it walks backward, identifying events that can generate it, such as "pressure rises" and "safety valve fails to open."

Any hazard identification technique can identify top events for FTA. Subsequent branches combine to generate the top event. Demands, equipment failures, common cause failures, human failures, or equipment unavailability compose the tree branches. Having frequencies and probabilities of the basic events, one can calculate the frequency of the top event through FTA. Events whose frequency or probability is available are basic events. They come from a database or other source.

TABLE 7.6
Checklist—Travel by Car

Item	Description	(C/NC)	Observation
1	Tire status	C	
2	Tire pressure	C	
3	Spare tire pressure	C	
4	Engine oil level	C	
5	Brake oil	C	
6	Brake operation	NC	To be checked
7	Cooling water level	C	
8	Air filter	C	
9	Petrol filter	C	
10	Gasoline	C	
12	Shock absorbers	C	
13	Wheel alignment	NC	To be provided
14	Tire balancing	NC	To be provided
15	Documents	C	
16	Driver's license	C	
18	Maintenance manual	C	
20	Windshield wiper	C	
21	Compulsory insurance	C	
22	Hand lantern	NC	To be provided
23	Trash bag	NC	To be provided

C, conforming; NC, nonconforming.

7.6.3 AUXILIARY TECHNIQUES

PRA, What-if, and Hazop identify the top events for FTA.

7.6.4 COMPLEMENTARY TECHNIQUES

ETA and CA assess the consequences of the top event.

7.6.5 FORMULARY

FTA has a specific symbology. Figure 7.1 shows the deployment of the top event *motor overheats*.

Assuming the frequency of voltage increase as 1.0/year and the probability of circuit breaker failure as 0.05, the frequency of excessive current will be 0.05/year. If the fuse is in a faulty state 1% (5%) of the time, a 0.01 *fraction dead time*, the excessive current in the motor occurs (0.05/year) (0.01) = 0.0005/year. If the frequency of primary failure of the motor is 0.01/year, the motor overheats 0.0005 + 0.01 = 0.0105/year or once every 95 years.

Risk Analysis Techniques

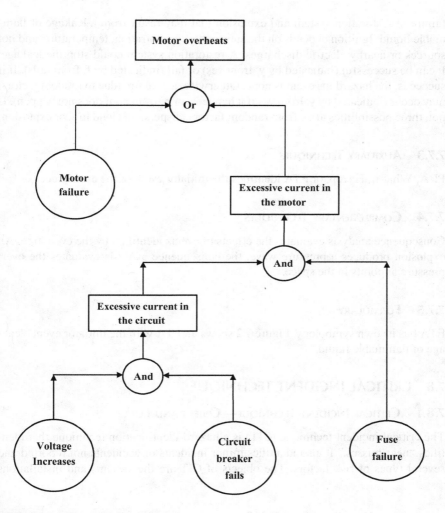

FIGURE 7.1 FTA—Motor overheating.

7.7 EVENT TREE ANALYSIS

7.7.1 ETA—Object and Focus

ETA or risk series is a hazard identification and risk analysis technique that identifies sequences of events that can follow a hazardous event. ETA may be qualitative or quantitative. Areas and emergency control systems contained therein are ETA's objects. The foci are the initiating event and the resulting events series.

7.7.2 ETA—Method

The top event of a fault tree is the initiator of an event tree. ETA is inductive (forward). Goes ahead, identifying events that may arise from the initiator, such as failure of restoration system and explosion that may result from a leakage of flammable liquid. Ignition depends on the substance, concentration, temperature, and hot sources or nearby electric discharges. A restoration system could stop the leakage. It can be successful (indicated by y, from yes) or fail (indicated by f, from fails). If it succeeds, the hazardous event is none. Otherwise, a pool fire (due to random factors) may occur (indicated by y, from *yes* if it happens; or n, from *no* if does not happen). If not, three possibilities arise from random factors: dispersion, cloud fire, or explosion.

7.7.3 Auxiliary Techniques

PRA, What-if, Hazop, or FTA identifies the initiator event of the event tree.

7.7.4 Complementary Techniques

Consequence analysis evaluates the effects of events identified by the event tree. An explosion produces a pressure wave; the consequence analysis evaluates the overpressure at points in the space.

7.7.5 Formulary

ETA has its own symbology. Figure 7.2 shows an ETA with the initiator event "leakage of flammable liquid."

7.8 CRITICAL INCIDENT TECHNIQUE

7.8.1 Critical Incident Technique—Object and Focus

The critical incident technique (CIT) is a hazard identification technique that identifies "near misses." It also identifies minor incidents or accidents not reported and several types of risk factors. The objects of CIT are the systems and installations

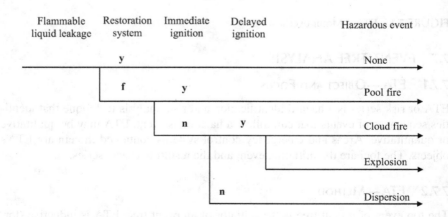

FIGURE 7.2 ETA—Flammable liquid leakage.

Risk Analysis Techniques

in the operational phase. The foci are events, attitudes, behaviors, conditions of the facilities, and interactions of persons with facilities and equipment.

7.8.2 CIT—Method

The method consists of interviewing people who work in the facility or system, obtaining the report of situations that almost produced accidents or manifestations of risk factors such as behaviors and attitudes.

7.8.3 Auxiliary Techniques

Interpersonal techniques gain people's trust, a key success factor.

7.8.4 Complementary Techniques

Planned inspection verifies the unsafe conditions identified by CIT.

7.8.5 Formulary

CIT demands a simple formulary to register the reported incidents, *near misses*, and unsafe conditions.

7.9 COMPARATIVE ANALYSIS

7.9.1 Comparative Analysis—Object and Focus

Comparative analysis compares the attributes of an object with patterns. The objects of the analysis are systems, facilities, and processes at any life cycle stage. The foci are the deviations from laws, technical standards, or particular requirements of an organization.

7.9.2 Comparative Analysis—Method

a. Compare the attributes of the object with standards, regulations, or other patterns.
b. Detect deviations.
c. Establish corrective or compensatory control measures.

7.9.3 Auxiliary Techniques

a. Planned inspection (operational stage).
b. Checklists ensure the implementation of corrective measures.

7.9.4 Complementary Techniques

Assessment of consequences establishes the priorities for correcting deviations.

7.9.5 FORMULARY

The formulary may be similar to a checklist.

7.10 INTERACTION MATRIX

7.10.1 INTERACTION MATRIX—OBJECT AND FOCUS

The matrix identifies hazards arisen from interactions between elements of the object. Objects are systems, facilities, processes, activities, and interventions. The foci are the interactions. The set of possible interactions contains the subset of the hazardous ones. Interaction matrix is a valuable tool to control the risks associated with interventions.

Some hazards result from interactions between three or more elements. However, some variables must assume a defined value or be within a particular range to generate a hazardous interaction. This happens in the fire triangle *fuel–oxygen–temperature* since the concentration of flammable substance in the air must be between the lower and upper flammability limits. Although those cases would require a three-dimensional matrix, the comparison in pairs leads to satisfactory results.

7.10.2 INTERACTION MATRIX—METHOD

1. List the elements of the object.
2. Use Table 7.7.
3. Analyze interactions (line-column crosses).

Interactions between construction materials and chemical substances, different chemical substances, and different activities are common in industrial maintenance.

TABLE 7.7
Interactions Matrix—Maintenance Services

A, B, C, etc., Are Activities, Substances, Material

Hazard: 0 = None, (1) = Low, (3) = Medium, (5) = High, and (?) = Unknown

	A	B	C	D	E	F	G	H	I	J		Interactive elements
A	5	3	3								A	Painting
	B	2	2								B	Welding
		C	4								C	Refractory
			D								D	Inspection
				E							E	
					F						F	
						G					G	
							H				H	
								I			I	
									J		J	

Risk Analysis Techniques

Ammonia and copper alloys or chlorides and stainless steel are harmful interactions for the metals. Interactions may occur between a drainage of flammable liquid and welding service; activities at different heights, but on the same vertical; industrial radiography and any other work in the same area; and two services performed concomitantly inside the same equipment.

The interaction matrix shows the importance of signage, order, and cleanliness as risk control measures. Order and signage reduce the probability of interactions. Cleaning reduces frequency of interactions and their severity. Disorder, uncleanness, and deficient signage cause accidents and failure of emergency control systems due to the high number of possible interactions. Combinatorial analysis helps to understand why. When four elements may interact, the number of possibilities is six. However, when ten elements may interact, the number of possibilities is 120.[2]

7.10.3 Auxiliary Techniques

a. What-if identifies hazardous interactions.
b. Models, maps, and drawings facilitate the risk control.

7.10.4 Complementary Techniques

FTA, ETA, and planned inspection can complement an interaction matrix.

7.10.5 Formulary

Table 7.7 shows a partially filled matrix. A risk analysis team identifies hazardous or undesirable interactions between maintenance activities inside an industrial equipment, assigning scores according to the hazard categories 0 (none), 1 (low), 3 (medium), and 5 (high). Categories 2 and 4 arise when a team member assigns one, while another assigns 3, or 3 and 5. The "(?)" indicates that the decision requires information not available, and the interaction needs additional analysis.

Painting (A) and welding (B) constitute a Category 5 hazardous interaction as it may result in a fire or explosion.

7.11 PLANNED INSPECTION

7.11.1 Planned Inspection—Object and Focus

Planned inspection observes an object, detecting deviations from standards. The objects may be installations, activities, systems, components, work environments, materials, and behaviors. The focuses are the deviations. Inspection is a sensor type I since it detects risk through its factors.

When direct observation is impracticable, the inspection uses the indirect one. The inspectors observe a secondary object and obtain the attributes of a primary object by inference. Observing the light emitted by a star, inspectors (astronomers) infer its composition; observing behaviors, inspectors infer attitudes.

Inspection can be an observation, experimentation, or test. An observation applies no stimulus. However, when it comes to people, it is impractical inspecting without stimulating. Experimentation applies a stimulus and observes the response or reaction of the object. Simulation is a type of experimentation that may involve the object, some components, or another system. The inspector infers a reality rarely available for observation.

7.11.2 Planned Inspection—Method

Planned inspection effectiveness depends on the reference standard, ability to detect deviations, and observation technique.

Standard is a benchmark for comparison. Some standards have parameters that we can accurately measure. Usually, workers exposure to noise is limited to 85 dB (A) per 8-h working day. If the measurement detects 90 dB (A), the deviation is 5 dB (A). Other patterns depend on nonquantifiable parameters. Consider housekeeping. Since we cannot estimate it quantitatively, we need to use a qualitative scale as *arranged*, *badly arranged*, and *well arranged*. However, *well arranged* for an observer may not be for others, since patterns depend on culture and are variable in time. An individual who enters a *badly arranged* place becomes shocked at first. Over time, accommodation occurs; the individual becomes accustomed to the environment and starts feeling in a normal place. However, it is possible to create patterns independent from people's minds, such as photographs, films, or standard installations. The inspectors may observe the standard, forming a vivid mental representation.

Ability to detect deviations can improve through training. An exercise improves the detection capability. The game of the seven errors is a crossword magazines pastime. Readers observe two figures apparently identical but containing seven differences, the seven errors. That game sharpens the ability to observe deviations. Inspectors can wake up their capacity for observation by playing that game before staring an inspection. Additional effects occur and they are as important as the expected enhancement of the capacity to detect deviations. The exercise promotes team relaxation, favoring communication and synergistic effect, removing the uncomfortable feeling for pointing defects on people's work.

Each system and activity requires a specific inspection technique. Piping welds require X-ray and gamma ray; electrical systems and thermography; and intracranial region, tomography, or magnetic resonance imaging. Workplace inspection also requires proper techniques. Let us take a closer look at work environment inspections, focusing on SOC (signage–order–cleanliness), aggressive agents, control systems, and behaviors.

An efficient technique uses Equation 6.1. With a checklist containing aggressive agents and types of aggressive energies, inspectors seek aggressive agents, immediate sources and contributors, aggressive energies, containment systems, rupture agents, possible targets, vulnerabilities, exposure possibilities, and emergency control systems.

Inspectors must identify risk promoters and inhibitors. SOC act as inhibitors of dangerous interactions and common cause failures. Therefore, any inspection must address those factors. Semiquantitative assessments of risk control system may use Table 6.8.

A single person may inspect a working place, but a multidisciplinary team, sharing different knowledges and experiences observe an object through different viewpoints. However, a group should constitute an organization, not an aggregate. Usually, the time available for development is short. Therefore, the leader must be competent and skillful on techniques presents herein, promoting the *cross-fertilization* and the holistic approach.

Communication with the personnel is a key factor for the effectiveness of the inspection. Personnel know most of the hazards, but they do not identify all credible hazardous events. The inspectors can help with concepts and techniques. Knowledge and experience coupled with concepts and techniques form a powerful hazard identification and risk analysis instrument. In addition to the safety aspects, inspectors must assess the organizational climate, culture, management system, and leadership. Inspections can be efficient instruments of risk control, spreading values, information, and knowledge on safety issues.

7.11.3 Auxiliary Techniques

CIT identifies hazards unrevealed through observation.

Checklist (pre-inspection) assures that the inspectors address all key aspects.

7.11.4 Complementary Techniques

Checklist (post-inspection) monitors the implementation of the recommendations of the inspection team.

7.11.5 Planned Inspection Formulary

The planned inspection demands a formulary to include the identified deviations and the recommended risk control measures.

7.12 ABNORMAL OCCURRENCES ANALYSIS

7.12.1 Abnormal Occurrences Analysis—Object and Focus

Abnormal occurrences analysis (AOA) addresses incidents, accidents, near misses, and any abnormal situation. The analysis identifies causes, consequences, and risk control measures. The objects are organizations, systems, or activities. The focus is everything that can present deviations from the normal state.

An abnormal occurrence is static when the deviation is a state. Inspections detect static abnormal occurrences. A photo can describe a static abnormal occurrence. An abnormal occurrence is dynamic when the deviations are events whose register requires a video and reports. The dynamic ones require reports because past scenes *evaporate*.

Generally, we express the risk through the expected or average value. The full description requires statistical distribution parameters (mean and variance). When we throw dices or roulettes, the results follow well-known distributions. Risks, however, are difficult to characterize because many factors participate, and their relations and interactions are complex. We can mention processes and equipment; people with their knowledge, skills, experience, creativity, and motivation; organizational culture, management system, organizational climate, and external factors.

Risk analysis detects risks before they become facts. Knowing the dice's characteristics (six equivalent faces), we infer that the probability of occurrence of each face is one-sixth. Similarly, knowing the risk factors, we infer their consequences.

Abnormal occurrence analysis detects the risk manifestations as real facts. The same way we observe the results of throwing a dice, and we infer the probability of a face to occur, we can analyze abnormal occurrences and infer the risk that generated them.

Abnormal occurrences that include incidents or minor accidents are more useful for inferring risks than those involving serious accidents, which have a lower frequency.

The effectiveness of the analysis of abnormal occurrences depends on its integration with other risk management instruments. In turn, the risk management system must reevaluate its instruments considering the occurrences.

7.12.2 AOA—Method

Registering without analyzing is impracticable. Every record contains some dose of analysis. Usually, an industrial plant has many occurrences classified as abnormal. Detailed analysis requires methods and techniques. Analyzing all of them with special methods or techniques consumes resources not always available. However, registering all occurrences and analyzing a selected sample produces good results because many records *speak for themselves*.

Several criteria can define which occurrences require deeper analysis. As parameters, we can choose risks, damages, or losses. Risk, or expected damage, is better than the actual damages since chance can produce results far from the average. Thus, an occurrence that does not produce damage can have a high-expected damage, as is the case of a falling object that, by chance, does not hit anyone. On the other hand, occurrences that cause serious damage can result from risks that do not require additional control measures, such as a person who falls when walking on a normal floor hitting the head against a wall. The expected damage associated with walking is much lower than that observed, and additional risk control measures are not practicable. It would not be feasible to cover all walls with shock absorbing material.

The method of analysis of abnormal occurrences is deductive. Starting from a top event, the occurrence, it identifies the causes. The basic concepts and the failures' theory are fundamental tools. The process has three steps:

1. Collecting information, visiting the scene, filming and taking pictures, interviewing people, and consulting technical data
2. Analyzing the occurrence
3. Concluding and recommending risk control measures.

Other techniques are of great value, such as planned inspection, What-if, FTA, ETA, and cause tree analysis (CTA). Specific forms containing fields to record history,

Risk Analysis Techniques

aggressive agents, harmful events, targets, damages, losses, hazards, and control measures are also useful.

7.13 CAUSE TREE ANALYSIS

CTA is a variant of FTA. The fundamental difference is that the FTA uses potential events, and the cause tree analysis uses facts. Therefore, the cause tree analysis is a type II sensor in the risk control system.

As we delve deeper into accident analysis, we realize that the most radical causes lie in the organization's regulatory function. Therefore, we can establish a direct link between that function and the damages and losses. Management system, organizational culture, and leadership constitute the regulatory function. These elements produce a force field acting on all the organization's resources. When they fail to promote the safety function, accidents occur. The causes related to management system come from the policy, guidelines, program structure, methodologies, and communication system; the culture's causes, from beliefs, ways of proceeding, myths, and values; and the leadership's causes, from directions, vision or conviction, posture, and behavior. Causes may also come from individual factors or external fields that the regulatory function failed to filter.

The mechanism of damages and losses contains five cause levels. The first level contains the facts; the second and third, the immediate causes; and the fourth and fifth, the root causes as indicated in Figure 7.3.

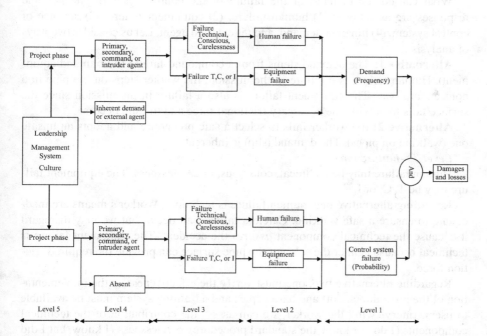

FIGURE 7.3 Damages and losses generation process.

Level 1—Action or no action

It is the most superficial level. It contains the immediate causes. It is the factual description. How did it happen? Answer: a demand occurred and the emergency control system failed.

Let us consider the abnormal occurrence: walking on a pipe, a worker slides, falls, and breaks a leg. The phase of collecting information revealed that his footwear had moist clay. It also ascertained that walking on pipes is a nonstandard practice.

When the worker walks, a potential energy difference exists between him and the ground. Therefore, the ground has latent aggressive energy. A containment system impedes the aggressive action. Worker's nerves, bones, and muscles, and the friction between footwear and the superficies of the pipe constitute the containment system. Moist clay acts as rupture agent. Demand is the worker's imbalance that breaks the containment and triggers the release of the aggressive agent (the ground) to act against the target (worker). In fact, the worker collides with the ground, but everything happens as if the ground collides with him.

The restoration system consists of the worker's ability to regain balance. That system is not effective. A safety cable would prevent the fall, but it is impracticable to use such a system for walking at ground level. It remains to count on protection systems, such as knee and elbow pads. This protection would avoid abrasions, but not the shock and twists. Therefore, there is no effective and practical emergency control system for this case, and once the demand occurs the damaging event occurs. Therefore, we must analyze why the demand occurs.

Level 2—Failure agent

What caused the demand or the failure of the control system? The possible responses are as follows: (1) human failure, (2) equipment failure, (3) absence of control system, (4) inherent demand, and (5) external agent. Let us consider two ways of analysis:

Alternative 1: The demand stems from a compound failure (human and equipment). Human failure is a failure in the mission; the worker steps on the pipe in a nonstandard way. The equipment failure is also a failure in the mission since the surface fails to provide the necessary frictional force and the slip occurs.

Alternative 2: the worker fails to select a safe procedure and adopts an unsafe one (walking on pipes). The demand (slip) is inherent.

Level 3—Failure type

Human failure may be technical, conscious, or carelessness. The equipment failure may be T, C, or I.

Regarding alternative one, human failure is technical. Worker's means are inadequate to ensure a safe walking. Careless failure can occur, but we may disregard it because the technical component *assures* the accident. The equipment failure is technical because neither the sole nor the pipe surface can provide the required friction force.

Regarding alternative two, one must verify the effectiveness of the implementation of the procedures. Soft and hard copies and a training system must be available to users. Interviewing the worker, we can assess the contribution of the technical components (I do not know the standard procedure), carelessness (I know, but I did not realize I was doing wrong), and conscious (I know, but I decided to adopt the alternative procedure). A poorly implemented procedure moves the causes to the technical side.

Level 4—Failure promoter

The failure promoter may be primary, secondary, command, or intrusive. One must also identify when the risk filtration failed and introduced the failure promoter.

Regarding alternative 1:

A secondary agent promoted the failure. The conditions were different from the design, and any other worker would fail.

Regarding alternative 2:

If the failure type is conscious, we must identify the promoters. The promoter is primary if the worker's history indicates that he disregards standard procedures. It is secondary if the time available is insufficient to execute the tasks unless workers take shortcut. The promoter is a command if supervisors close their eyes and pretend that they do not see unsafe procedures or when accepting them as common practices. An intrusive promoter is present if people outside the system give opinion about the smart way to do the work.

If the failure type is carelessness, we may identify a primary promoter if the worker is absentminded. A secondary promoter may be the cause if the working routine has many tasks, and the frequency of each one is low. A command promoter would be the cause if supervisors are confused and give many orders and counter-orders. Finally, an intrusive promoter is in place if supervisors outside the system request the worker to perform tasks not included in its routine.

If the failure type is technical, we may ascertain a primary promoter if the other workers know the procedures. A secondary one may be the cause if procedures are unavailable or workers have no training. A command promoter is the probable cause if the procedures are unclear, or supervisors do not give instructions. The intrusive promoter is present when people outside the system influence the work practices. Intruders never act alone. Workers listen to intruders when good procedures are not available.

Level 5—Root cause

Why was the failure promoter successful?

The answer is as follows: *The regulatory function of the organization failed.*

NOTES

1. The word Hazop derives from hazard + operability.
2. The number of possible combinations between n elements taken k by k: $C_{n,k} = n!/[k!(n-k)!]$, where $n! = n(n-1)(n-2)(n-3) \ldots 3.2.1$.

BIBLIOGRAPHY

1. Lewis, Elmer Eugene. 1987. *Introduction to Reliability Engineer.* USA: John Wiley & Sons.
2. Lees, Frank P. 1996. *Loss Prevention in the Process Industries.* UK: Reed Education and Professional Publishing.
3. Kletz, Trevor A. 1985. *An Engineer's View of Human Error.* UK: Institution of Chemical Engineers.

4. Macdonald, David. 2004. *Practical Hazops, Trips and Alarms*. UK: Newness.
5. Kletz, Trevor A. 1999. *What Went Wrong?* USA: GPP.
6. Kletz, Trevor A. 2001. *Learning from Accidents*. UK: GPP.
7. Roland, Harold E. and Moriarty, Brian. 1990. *System Safety Engineering and Management*. USA: Wiley Inter-Science.

8 Value Analysis in Safety

Will a war always be necessary to motivate us to find such rational alternatives?

Harry Erlicher, vice president at General Electric Co.—GE

8.1 INTEGRATING VALUE AND RISK ANALYSIS

Value analysis emerged after World War II to rationalize the usage of raw materials. The fundamental concept is that of function. Its purpose is to increase the value of objects by executing their functions at lower costs. "Value engineering" emerged to designate the development of new products, maintaining the term value analysis for the existing ones.

Value analysis promotes hazard identification through the concept of aggressive function. Moreover, it provides a method to select effective and low-cost risk control measures, promoting the integration of the safety function with other vital functions of the organization.

8.2 AGGRESSIVE FUNCTION

Risk analysis emerged from the military area. Value analysis is a civilian creation induced by the limitation of materials during World War II.

What does this do? The fundamental question of value analysis identifies functions.

What can go wrong? The fundamental question of risk analysis identifies hazardous events.

What does this wrong? The answer is an aggressive function linking value and risk analysis. Needs, quality, and satisfaction give way to vulnerabilities, hazards, and aggression. Instead of specifications, we have tolerances. For example, making noise is an aggressive function. Those who purchase a household appliance do not need it. On the contrary, they are vulnerable to noise. In some cases, high-utility products became extinct because of aggressive functions. A notable example is an oil for electric transformers, known commercially as *Askarel*, destroyed all over the world by incineration and no longer produced.

Figure 8.1 presents the common elements of value analysis and risk analysis: systematics, creativity, and teamwork.

Value analysis adds the functional analysis to safety studies. Failure modes and effects analysis identifies "what can go wrong." Hazard and operability studies (Hazop) detect hazardous deviations. Aggressive functions reveal hazards.

FIGURE 8.1 Common elements of value analysis and risk analysis.

8.3 BASIC CONCEPTS OF VALUE ANALYSIS

8.3.1 OBJECT OF VALUE ANALYSIS

Everything that presents quality and cost concerns can be the object of value analysis.

8.3.2 FUNCTION

Function is the fundamental concept. It is everything the object does, such as emit sound, contain volume, mark hours, and reflect light. A state verb (to be) does not describe a function but expresses state or condition, intrinsic characteristic or specification, such as being round or transparent.

To perform functions, some objects require stimuli, such as energy or information. Thus, a lamp emits light when receives electrical energy, bodies emit light (images) when light falls on them, and people leave a place when an alarm sounds. The prospect of satisfying needs is a motive for action (motivation) and encourages people to perform functions. Some functions generate their own stimuli. In combustion, the function "release heat" generates its stimulus (temperature). Therefore, when an object fails to perform the expected function, or it is in a fault state, or it does not receive the adequate stimulus.

We describe a function through an active verb and a noun. Action (verb), direct object (what), performance (how much), reliability, indirect object (for whom, whom), and constraints (time, space) are functions attributes. Electric motors "supply power," a function with the following attributes: action, provide; direct object, power; performance, 100 HP; reliability, 99.9% after 10,000 h, operating at 440 volts, at a maximum ambient temperature of 40°C and no dust.

8.3.3 QUALITY

Let us consider a person (client) who needs to warm food. The stove *burns gas*. Burning the gas is what the stove does, and its quality results from how much that function satisfies the need. However, it can *burn people*, an aggressive function and a negative side of *quality*.

Value Analysis in Safety

8.3.4 Price

Price is what we give in exchange for something. In a way, it represents a loss. Let us establish a broad concept, including any physical, mental, or moral effort in the price. Therefore, the price includes the effort to seek an object, the time we wait to receive it, and its aggressive functions. The price we pay to enjoy the benefits of the equipment must include the noise among other aggressive functions.

8.3.5 Absolute Value

The absolute value stems from the quality-to-price ratio.

$$\text{Value} = (\text{Quality}+)/\text{Price}, \text{ or } V = Q/P \tag{8.1}$$

Quality+ is the positive side of quality. Price includes the negative side. The higher the quality, the higher the absolute value; and the higher the price, the lower the absolute value. When the price tends to zero, the value tends to infinity meaning that it is very high whatever the quality. However, a cheap product may hide a price that one pays latter through damages and losses.

Analysis is the division of a whole into parts and a detailed study of these parts. Because quality and price derive from functions and costs, value analysis is a thorough study of functions and costs.

8.3.6 Relative Value

Let V_1 be the absolute value of an object chosen as reference and V_2 be the absolute value of a second. The relative value of the second object is

$$V_{21} = V_2/V_1 \tag{8.2}$$

The relative value of the first object is one.

$$\text{As } V_1 = Q_1/P_1 \text{ and } V_2 = Q_2/P_2, \; V_{21} = (Q_2/Q_1)/(P_2/P_1) \text{ or } V_{21} = Q_r/P_r, \tag{8.3}$$

where Q_r is the relative quality and P_r is the relative price.

The client acquires the object of the highest relative value. Consider a product, unique in the market. To determine the relative value, we must consider that V_1, the reference, is null since what does not exist has no value. Therefore, mathematically, V_{21} would be infinite. In the absence of competition, the relative value raises, and the customer agrees to pay a high price. However, if the price goes up, the absolute value falls, the customer seeks alternatives, and when competitors appear, the relative value suffers a vertiginous drop.

8.3.7 Clients of the Organization

Clients' satisfaction determines the quality of an organization. Client usually means consumer. Providing a product to that client is the basic function of the organization.

The quality-to-price ratio determines the value of the product. However, the organization has other clients.

Client–sponsor or shareholder (in companies) is who invests monetary resources. They have a basic need, to ensure his capital profitability. An organization meets that need by *providing profits* or returns. The sponsor pays a price, the immobilization of resources, and the risk of loss. Other investment alternatives will determine the relative value.

Client component or employee (in companies) is who puts their human resources in the organization. His basic need is to support the family. An organization meets this need by *paying wages*. Price is the time they work and the exposure to occupational risks. Other companies that could contract them determine the relative value.

The client community is who invests environmental resources and needs to build schools, health posts, and other social works. An organization meets their needs by *paying taxes*. Price is the risk to the health, properties, and the environment. The installation of other organizations will determine the relative value.

Figure 8.2 represents the holistic value of an organization, resulting from the relationships with all clients. For example, the consumer may see a pulp mill as a high-value organization, while the community considers it as a low-value one due to the emission of pollutants.

Referring to needs, we usually think about the basic or physiological ones. However, people have a set of needs (physiological, safety, association, esteem, and self-realization) that value analysis must consider.

8.4 FUNCTIONS CLASSIFICATION

8.4.1 Basic—Complementary and Supplementary Functions

A basic or main function is the reason for the existence of an object. When the object is an organization, the basic function is the mission. The basic function of a watch is *to set hours*. The mission of a factory of soap is *to produce soaps*.

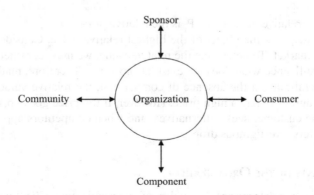

FIGURE 8.2 Holistic value of an organization.

Value Analysis in Safety

The object may perform complementary functions in parallel with the basic one, completing it under one or more than one point of view. *Illuminate display, resist impact,* and *resist pressure* are complementary functions of a watch. In some cases, the complementary functions are vital functions.

Supplementary functions are parallel and independent of the basic, such as *count seconds* and *show appearance* in a watch. In many cases, a supplementary function is more important than the basic one. Generally, this occurs with esteem functions.

8.4.2 Auxiliary Function

Any top-level function (basic, complementary, or supplementary) results from a series of auxiliary functions. As we move toward the necessity by answering the "why" question, the level of a function increases. The resistor of an electric shower performs the function "resist electric current," necessary to produce the higher level function *release heat*. The function *warm water* is at a higher level, closer to *take a warm shower*.

8.4.3 Use Function

A use function meets basic needs. A watch executes *set hours* and *count seconds*.

8.4.4 Esteem Function

Esteem functions give prestige to objects such as *provide appearance* of a watch. In many cases, an esteem function is more important than the basic one.

8.4.5 Necessary Function

A necessary function satisfies directly or indirectly some need. Indirect satisfaction occurs when the necessary function is auxiliary of another function.

8.4.6 Unnecessary Function

Unnecessary functions do not satisfy needs. Value analysis identifies and eliminates them. They result from tradition, convenience, or copying of previous projects without critical analysis.

8.4.7 Support and Collateral Functions

Some functions (support function) generate associated functions (collateral). The inherent collateral function necessarily follows the support function. The circumstantial collateral function follows the support sporadically. *Produce voltaic arc* generates *emit ultraviolet rays*, an inherent collateral function; *analyze gases* may generate the function *release toxic gas*, a circumstantial collateral. A collateral function may be useful, neutral, or aggressive.

8.4.8 Aggressive Function

An aggressive function can cause damage. It may be a useful function, including the basic (*emit X-rays*, the basic function of an X-ray apparatus), an inherent collateral (*emit ultraviolet rays*, inherent in producing voltage arc), or a circumstantial collateral (*release toxic gas*, circumstantial of *analyze gases*). Identifying aggressive functions, we must identify the vulnerabilities of the targets to those functions.

8.4.9 Real Function

Real function is what the object really does. A cigarette *emits dioxins, smoke, and carbon monoxide*.

8.4.10 Fictitious or Imaginary Function

The object does not do an imaginary or fictitious function, but the client imagines or supposes that it does. It can be a useful or aggressive function. For some smokers, the cigarette *improves image*.

8.4.11 Perceived Function

A perceived function is what the client "sees." It can be real or imaginary. Image and quality result from perceived functions. In many cases, advertising emphasizes real useful functions, hides the aggressive ones, and suggests functions that the product does not execute.

8.4.12 Revealed and Undisclosed Functions

Revealed functions are the ones we know the object performs or can perform. We can detect *emit noise*, *emit light*, and *emit odor*. However, some functions remain hidden; the object does something, and we do not know that. They may be permanent functions, but we do not detect them due to lack of information, knowledge, methods, or technology. Radioactive materials perform the function *emit gamma rays*, but no one knew it until the discovery of radioactivity. In other cases, the object performs the function only under certain circumstances or conditions. Some practical rules promote the identification:

   a. Considering the entire life cycle of the object. Nonpolluting products may emit pollutants during fabrication (cellulose) or when disposed into the environment (plastics, tires).
   b. Considering other operating conditions, such as *emit toxic gas* when burned (plastics, some food) and *emit carcinogenic fibers* when rubbed (asbestos).
   c. Considering the hazards of new products, identifying toxicity and biodegradation functions. The creators of the *Askarel* did not know its highly aggressive functions.
   d. Consider when and where, such as emitting light *when heated* and *release ink in the water*.

Value Analysis in Safety

8.4.13 VITAL FUNCTION

A function is vital when the object may become extinct without it. The basic function is vital, but complementary functions may be vital too. *Preserve the environment* is vital for businesses and products.

8.4.14 PASSIVE AND ACTIVE FUNCTIONS

Passive functions involve reaction, such as *resist stress* and *hold volume*. Active functions involve initiative, such as *provide heat* and *emit noise*.

8.5 FUNCTIONAL DIAGRAM

A classical method for describing functions generates a list of functions with their descriptions. It is satisfactory for simple objects or simplified analyses. However, the list does not provide a picture of the system and its process. The set of functions forms aggregates without the relationships. Another deficiency arises when one establishes the functions' hierarchy through the functional numerical matrix that compares functions according to an attribute as presented in Table 8.1.

In turn, functional diagramming describes functions and establishes their relationships in serial and parallel arrangement, providing a clear view of the systems and their problems, improving the evaluation, questioning, and the search for alternatives. Facilitating the identification of functions related to safety, environmental preservation, and human development, it extends the value analysis beyond the producer–consumer relationship. It also improves the selection of target functions for questioning.

In a diagram, functions hold two basic types of relation: series and parallel. We can develop a sequence by unfolding any function. A sequence represents a process that can be real or virtual. In the virtual process, the unfolding is in the mind. Actually, the action is unique. Figure 8.3 shows a virtual process of a nameplate

---

TABLE 8.1
Functional Numerical Matrix

A, B, C ... Are Functions

Relative Importance: 0 = Nonexistent, 1 = Low, 3 = Medium, 5 = High

A	B	C	D	E	F	Function	Score	Relative Importance (%)
A	A_2	A_5	A_3	E_3	A_4	A	14	31
	B	B_3	D_1	E_5	B_2	B	5	11
		C	D_1	E_5	F_1	C	0	0
			D	E_4	D_1	D	3	7
				E	E_5	E	22	49
						F	1	2
						Total	45	

FIGURE 8.3 Serial functions—virtual process.

fixation. The functions are *to fix* and *to screw*. Fixing is the why and screwing is the how. *Fix the plate* is at a higher level than *screw the plate*.

Two or more actions compose an actual process. Figure 8.4 shows the actual process with the functions *to drill* and *to screw*. Drilling is a step necessary for screwing.

To develop the series by moving to the left (solution), we make the question (how?). To develop to the right (objective, necessity) we make the question "*why?*" In the real process, the question "*for what?*" is more adequate than the question "*why?*" We can also use the nine questions proposed by Bytheway[1] to construct the diagram. In the nameplate case, a Bytheway's question is as follows: What do you really want to do when you screw the nameplate? The answer is *to fix nameplate*. Applied on *fix nameplate*, the answer is *identify the manufacturer*.

The parallel functions may have an *and* or an *or* relationship. In the *and*, all parallel functions must work to produce the top-level function. Figure 8.5 shows parallel functions producing a top-level function through the *and* relation. Any lower level function arises responding to the question "what is required for (upper level function)?" Asking *is it necessary*? the answer is *yes*; asking *is it sufficient*? the answer is *no*.

In the *or* relationship, any function is sufficient to produce a higher level one. Therefore, a function may be alternative to the other. However, alone they may fail to assure adequate performance, and we need to maintain two or more parallel

FIGURE 8.4 Serial functions—real process.

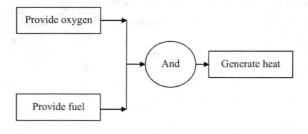

FIGURE 8.5 Parallel functions—real process.

Value Analysis in Safety

functions. Figure 8.6 shows parallel functions producing the top level through an *or* relationship.

A series may have the general appearance presented in Figure 8.7, with serial, parallel, and collateral functions.

8.6 COMPARING FUNCTIONS

Table 8.1 presents the Functional Numerical Matrix used to compare functions according to some parameter. Let us consider the *importance level*.

Comparing functions A with B, we place the winner at the junction row A/column B and assign a score one (low), three (medium), or five (high) to indicate the difference between their importance. When a team performs the analysis, the scores two and four indicate lack of consensus (between one and three or three and five). Table 8.1 shows the function E as the final winner with 49%.

8.7 HOLISTIC DIAGRAM

Let us present a holistic diagram assuming the following:

a. Every system has a basic function, the mission, the reason for its existence.
b. The organization must perform complementary functions essential to its survival.

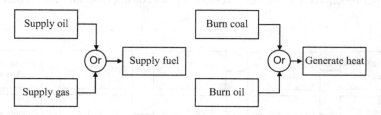

FIGURE 8.6 Parallel functions—*or* relationship.

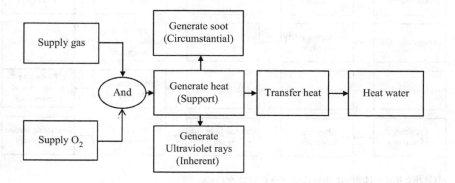

FIGURE 8.7 Parallel, serial, and collateral functions.

(Promoting) productivity, (promoting) quality, (promoting) safety, (promoting) environmental preservation, and (promoting) human development are the vital functions. They interrelate, interact, interdependent, and present a significant positive synergy. Neglecting one of them compromises the success of the organization. Therefore, the organization must treat them at the same level of importance.

Productivity is (quantity produced)/cost, regardless of quality. To analyze costs we must consider the resources involved. Quality refers to products and services intended for the consumer. Safety covers the components of the organization, the community and the consumer. Human development encompasses the four clients and means fostering conditions for advancement in the *Maslow*'s scale. Environmental preservation encompasses flora, fauna, water medium, atmospheric air, anthropic environment, and soil. A holistic diagram assumes the aspect shown in Figure 8.8. Functions that appear in more than one series demand special attention because they contribute for more than one vital function.

8.8 QUESTIONING FUNCTIONS

Innovations, problem solutions, and improvements arise from questioning functions. We choose functions as targets for questioning according to some criteria.

Usually, value analysis selects critical functions, defined as those with a high cost/importance ratio, limiting the scope to the producer–consumer relationship. However, in many cases, we have other reasons to question functions. In general, the target functions have negative results. Let us consider the mission, productivity,

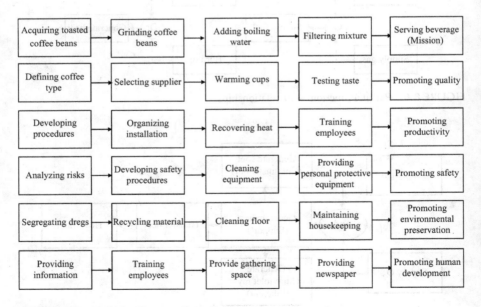

FIGURE 8.8 Holistic diagram of a coffee service.

Value Analysis in Safety

quality, safety, environment, and human development as the parameters to select target functions. Therefore, the following functions are potential targets:

1. The mission, since any modification in this function, affects strongly the overall system.
2. Functions of the productivity series that exhibit high failure rate or low performance and functions of other series that have a negative impact on productivity. We must know the relative contribution of each series to the total cost of the system. For this, we need express resources in monetary terms, which can be done with greater or less precision.
3. Functions of the quality series that exhibit high failure rate or low performance and functions of other series that have a negative impact on quality.
4. Safety functions that present high failure rate or low performance and functions of other series that are aggressive or have aggressive functions associated in an inherent or circumstantial way. In the same way that the cost defines the target functions in the productivity criterion, the associated risk defines target functions in the safety criterion. We must consider the risks to persons, environment, and property that may manifest themselves at any stage of the life cycle of the object.
5. Environment functions that present high failure rate or low performance and the functions of other series that cause or can cause significant environmental impacts.
6. Human development functions that present low performance.

An effective questioning combines systematic and creative approaches. We may question the functions in five depth levels: need, concept (principle), conformation (form), solution (means), and procedure (how to obtain the function). In each level, a creative method generates alternatives to perform the function. This systematic method is more productive than a loose brainstorming. The questioning addresses five levels. At each level, it is fundamental to use creativity techniques to generate ideas that, once selected, produce viable alternatives.

1. Necessity is the deepest level and refers to the reason for the existence of the function. If the function is not necessary, we can disregard the subsequent levels. The first question is "is the function necessary?" If the answer is no, the analysis ends. If the answer is yes, we no longer have the freedom to question the need, but we can question the concept.
2. The concept level refers to the basic science involved (mechanical, optical, or thermal printer can record characters). A second question is "could we use another concept to perform the function?" (Replacement of piston engines by jet turbines and the thermal and mechanical printers by the laser one are striking changes of concept).
3. Conformation level refers to the basic design or architecture (a staircase may be a spiral, fire escape, or escalator type).

4. Solution level involves engineering, detailed design, materials selection, and layout (the ladder's material may be wood, aluminum, and iron).
5. Procedure level is how to do, obtain, mount, or install the object that will perform the function (we can buy a ladder or manufacture it).

Let us analyze the function *to access a locking valve* to isolate a pipe as target for questioning. Currently, a man uses a ladder to access the valve.

1. Necessity

    Is it necessary *to access the valve*? Why arrive at that place? Are there alternatives? Is there another procedure that eliminates the function? If the answer is "It is necessary," we jump to the concept level.

2. Concept (principle, basic science)

    Could we use another concept (principle) to access the valve? The team could suggest an electrical alternative, such as a motorized valve remotely controlled or an elevator. However, if the answer is *the concept must be as designed*, we jump to the conformation level.

3. Conformation (basic design, architecture)

    Could we use another form for the ladder? The team could suggest a helicoidally stair. However, if the answer is *the form must be as designed*, we jump to the solution level.

4. Solution (means, detail design, engineering)

    Could we use a different material or change the thickness? The team could suggest iron or wood. However, if the answer is *the material and specifications must be as designed*, we jump to the procedure level.

5. Procedure (how to obtain the object that will perform the function)

    Could we use another procedure to obtain the ladder? The team could suggest buying it, making in house, or hiring the execution.

NOTE

1. Charles Bytheway, 1967.

BIBLIOGRAPHY

1. Cardella, Benedito. 2011. *Value Analysis: A Holist Approach*. Sao Paulo: Scortecci.
2. Cardella, Benedito and Ramalho, Andre. 2007. Value engineering focused on HSE: The application of value engineering on HSE management. *Kuwait: The Fourth International Health, Safety, Environment and Loss Prevention*, American Society of Safety Engineers: Kuwait.
3. Miles, Lawrence D. 1989. *Techniques of Value Analysis and Engineering*. USA: Lawrence D. Miles Value Foundation.

9 Human Failures

When I go to a country, I do not ask if that country has good or bad laws, since good and bad laws every country has, but if they enforce their laws

Montesquieu

9.1 FAILURE ANALYSIS

When a component performs a function improperly or fails to perform it, we say that it fails. The component may be a person or equipment. Person and equipment are systems' active elements. Failures are risk factors and, in almost all cases, accidents occur due to some type of failure. Much of the risk control function is to identify potential failures and take measures to eliminate them, reducing their frequency or neutralizing the effects. Therefore, the failure control is a part of the risk control. Understanding the failure mechanism is essential for the analysis of abnormal occurrences. Failure analysis includes identifying the failure mode and type, promoters and inhibitors, and the life cycle phase that introduced the failure and its promoters.

Any component, person or equipment, can fail in five ways:

1. Omission, when they do not perform or partially perform an intervention, task, function, or step.
2. Mission, when they perform an intervention, task, function, or step incorrectly.
3. Strange act, when executing an intervention, task, function, or step that they should not do.
4. Sequential, when executing an intervention, task, function, or step in the wrong sequence.
5. Temporal, when executing an intervention, task, or function at the wrong time.

Human failure is a notable one because almost all failures have some human factor contribution. For this reason, we only address equipment failures to the extent that we can establish their relationships and make analogies with human failures.

Two different situations require failure analysis. The first is the post-emergency, when everything that could occur has already occurred and the analyst describes failures, identifies causes, and assesses the effectiveness of the failure detection and interventions. Generally, the analyst was not part of the system nor participated in the events. The second is the emergency, the analyst being part of the system as operator, pilot, or doctor. Failures are still occurring; aggressive agents are acting and the emergency control must stop them. Signs and symptoms are available, but the correct intervention depends on the identification of the causes. Some people have great performance analyzing the post-emergency but paralyze during emergencies. In either situation, training in failures' theory is essential for a good performance. Knowing the failure modes is essential to execute failure modes and effects analysis (FMEA).

Chapter 7 addresses post-emergency through the analysis of abnormal occurrences. Let us present some remarkable situations.

In the first situation, we have an indicator and infer the cause that may be a failure or the action of an aggressive agent. However, no confirmation is available. One must consider that an indicator is a fact, and from it, we infer the occurrence of other facts. However, more than one fact may generate the same indicator. If our inference is correct, other indicators must assume consistent values. Let us consider an operator observing the control panel of a crude distillation unit when he sees the indication of the load flow meter of a fired heater dropping to zero. Hastily, he concludes that the load pump stopped working. Immediately, he shuts down the plant but soon realizes that the pump had not stopped. Actually, the flow indication dropped to zero because of a register's failure. From the flow indication, he inferred pump's failure. However, if the pump had gone out of operation, the pressure indicator (another indicator) would have assumed a very low value. In another case, at the opposite extreme, instruments indicate that a dangerous event is in progress, and the individual, in an unconscious defense process, does not believe that it could be true. Therefore, a trained operator must observe a change, select a cause, and verify that other expected effects from such cause are also occurring. Training in fault detection must habilitate operators to know the effects of faults on the system and on other components and, from those facts, infer the causes.

In the second situation, we have effects and find an abnormality that could produce them. Let us assume that a confirmation is available. However, we must check whether other abnormalities could produce the same effects. Moreover, they might be even more serious than the first one. One must consider that a complex system has probabilistic cause–effect relations and other correlations. Therefore, great care is necessary when concluding that the abnormality is the unique cause of observed effects. The natural tendency is to stop the investigation when detecting a cause because the feeling of success dissipates the tension generated by the unknown. Satisfied with the finding, an individual allows the emergency progressing to dangerous stages. Consider the following cases:

a. A doctor examines the patient who has a fever and other signs, finds red tonsils, diagnoses tonsillitis, and fails to detect a thyroid inflammation.
b. Examining a patient who had vomited and presented other complaints, a doctor finds a stomach ulcer and fails to detect the constriction of an intestinal hernia that requires urgent intervention.
c. Searching for the cause of a contamination, an operator of a chemical plant finds a block valve poorly closed and fails to detect other sources.

In the third situation, an emergency starts and personnel must predict its evolution. Some cases escalate fast; others have a stabilized period and suddenly escalate into a

Human Failures

catastrophic hazardous event. A striking case occurred during a fire in a fuel oil tank (Venezuela, December 19, 1982) which evolved into a boilover,[1] killing 150 people, injuring an unknown number and causing 50-million-dollar damages. The event tree analysis is a useful technique when dealing with these situations.

9.2 TECHNICAL FAILURE

Failure is technical when the cause is lack or inadequacy of resources. A person fails because does not know how to do or cannot do it right. A technical failure is most likely to keep repeating if the conditions remain the same.

Technical failures may be random or systematic. When the failure is random, the results disperse without deviation from the desired value that coincides with their mean. When the failure is systematic, the mean shows a deviation from the desired value.

Examples of technical failure are as follows:

1. A worker tries to separate green objects from the red ones but fails. He is color-blind. Inappropriate resource: man (biological).
2. After an alarm sound, an operator has 10 s to shut down a pump. He delays and the equipment damages. An analysis demonstrates that no one could act before 20 s. Inappropriate resource: time.
3. A fire injures many people because the evacuation corridors are narrow. Inappropriate resource: space.
4. A combat group failed to control a fire because the staff tired quickly despite their great knowledge and experience. Inappropriate resource: energy.
5. A worker performs a weld with several defects. The electrode specification is inadequate. Inappropriate resource: material.
6. A worker fails to control the measurements of a piece. The required accuracy is 0.1 mm, but he uses a 0.5-mm-accuracy apparatus. Inappropriate resource: equipment.
7. A man crosses the street when the pedestrian traffic light is red. He had never been in a city. Inappropriate resource: information.
8. A tank's level indication exceeds 100%, and the operator increases the pump flow. A pump goes into cavitation. The indication was false, and the level was actually low. Inappropriate resource: information.
9. A homemaker prepares a cake and puts too much yeast. The cookbook contains an error. Inappropriate resource: knowledge.
10. An operator opens valve A when the correct action would be to open valve B. He had received instructions from the supervisor to open valve A. Inadequate resource: knowledge.
11. A project team fails to identify project's hazards. They do not use a proper hazard identification technique. Inappropriate resource: procedure (risk analysis methods).
12. A driver causes a collision. He received a driver's license a week ago. Inappropriate resource: ability (or skills).

Comparing human and equipment behaviors, we can find some similarities regarding failures. Let us call failure T (technical), which occurs when the equipment cannot perform the required mission. This can occur when

1. The function is not in the equipment design.
2. Working conditions are different from design (material, energy, or environment).
3. The equipment is in a faulty state.

Any equipment is prone to failure T. Equipment that performs static functions, such as *contain volume* and *resist stress*, fails only on T type. Examples are pipe rupture, pressure vessel leakage, beam, or screw rupture. Equipment that performs dynamic functions, such as a centrifugal pump, may fail on T type when demanded to operate beyond the design envelope (designed to pump water but pumping a very viscous fluid).

9.3 CARELESS FAILURE

9.3.1 Carelessness Failure Characterization

Carelessness, inadvertent, or unconscious failures stem from the inability of unconscious and automatic human mechanisms to control man's actions. The individual has all the resources and acts correctly a great number of times, but sporadically he fails. This is the characteristic of the carelessness failure. It is sporadic, has a small probability of occurrence. However, far from making it less dangerous, this makes it a treacherous risk factor, generating perplexity and no effective control measures. Consider the following cases:

1. A man puts shaving cream instead of toothpaste on the toothbrush. He had performed this operation correctly hundreds of times.
2. A driver passes through an intersection when the traffic light turns red. He is an experienced driver but only after crossing, he realizes what he did.
3. A house cleaner forgets to turn off the iron and causes a fire. She usually turns off the iron shortly before finishing ironing the clothes.
4. An accountant fills a check incorrectly. He had filled hundreds of checks before.

Carelessness failures can lead us to regard man as a danger to systems, facilities, processes, and his own fellows. However, the human being is an unbeatable problem eliminator, planner, and decision-makers. His weakness arises when he acts as a component, responding to specific, scheduled, and repetitive stimuli. He fails more than machines, but the reason is his own complexity. Man can perform a task in many ways. However, few or only one is safe. Machines use few or even only one way. Therefore, they can always do the safe way.

Machines sophistication makes them similar to man. They can take decisions on complex matters, but may also commit complex failures. Replacing simple but

Human Failures

safe electromechanical devices with sophisticated electronic ones increases the complexity. Electromechanical systems contain simple cause–effect or stimulus–response relations. The sophisticated electronic ones have complex relations, similar to humans. Moreover, when sophisticated systems fail, human intervention requires much more knowledge and skills.

Low-tension situations generate inattentiveness; high-tension situations promote confusion and stress; both amplify carelessness failures.

9.3.2 Inattention and Confusion

Routine situations generate inattention. Human inherent difficulty sustaining attention continuously for long period is the basic cause of carelessness. Therefore, the more the system depends on that attention to operate safely, the higher the probability of failure. Continuous monitoring and repetitive tasks contribute to attention deterioration. During continuous monitoring, attention tends to deteriorate rapidly after half an hour, demanding task rotation.

Emergencies generate confusion. Reversion to the stereotype of the original population is common during emergencies. Even if the most recent training has been in the opposite direction, the individual reverts to his population's stereotyped response. If light is "on" when the switch position is "up," but in his population's standard the light is "on" when the switch is "down," during emergency he probably will try to put the switch "down" to turn "on" the light. Therefore, human factors' engineering must take great care to prevent stereotype violation when designing instrumentation and control systems.

9.3.3 Handling Carelessness Failures

These failures are difficult to prevent because they rarely show observable patterns. Some authors consider that inadvertent failures are unpredictable. However, we can predict what will occur if we store poison along with food.

Careful design of man–machine interface can minimize carelessness failures. The color, shape, and other means can distinguish instruments and controls, minimizing confusion. A well-designed workplace with comfortable seating, adequate lighting, temperature, and humidity control contributes to reducing carelessness failures.

When the human being acts as a component of a system in a programmed and repetitive way, he presents a probability of failure little affected by education, training, motivation, praise, or punishment. To reduce carelessness failures we must recognize the human being's characteristics. We cannot change the human nature, but we can modify projects and procedures, adapting them to man. Based on that principle, we eliminate the possibilities of failure modifying design, working conditions, and equipment.

We can reduce carelessness failures introducing control measures in the design and operational phases. In the design phase, we can install devices or foolproof mechanisms such as the light signal indicating hand brake set in the automobile dashboard and a concrete wall preventing trucks from hitting buildings when parking. In the operational phase, selecting criteria, work routine, stress level, training, and good work practices.

Selection. Human performance varies from person to person. If the selected person features job requirements, the failure's probability decreases.

Work routine. The ability to maintain attention drops dramatically over time. Constant surveillance tasks require a rotation scheme. In World War II, the British discovered that the maximum time that a lookout of submarines could stay at the observation center was about half an hour. After this period, the probability of perceiving an enemy ship or aircraft was unacceptably low. The consciousness of the danger was insufficient to maintain him attentive.

Psychological strain level. Psychological strain alters human performance. Under low strain, persons neglect their duties. Strain increments raise performance, but from a certain level on, any new increment causes drops. Emergency increases strain and turns performance extremely low.

Training. Training effectively reduces inadvertent failures during emergency.

Good working practices. Work practices reduce the likelihood of harmful events. Practices are non-written procedures. Therefore, they are between the management system and organizational culture. Washing hands before meals, separating food and chemicals, checking inside shoes to prevent venomous animals are good practices.

9.3.4 "Inadvertent" Equipment Failures

By analogy with human carelessness or inadvertent failure, let us call "Failure I," which occurs when equipment is capable of performing the function with good performance most of the time; however, it fails sporadically. This can occur in the following cases:

1. Equipment has components whose fault state manifest sporadically (bad electrical contact, for example).
2. Some work condition varies sporadically (improper material, impurity).

Failure "I" occurs in equipment that performs signal transmission functions such as wires, transistors, and electronic valves.

9.4 CONSCIOUS FAILURE

9.4.1 Conscious Failure Characterization

The failure is conscious when triggered by alternative procedures that involve greater risks than the standards. Alternative procedures aim to achieve other goals and interests as cost, schedule, productivity, quality, comfort, or status. The person knows the safe procedure, established by standards, but deviates from it, not by carelessness, but by conscious decision to do so.

As conscious failures result from violation of rules and standards, we may apply a thought of the philosopher Montesquieu on accident prevention: "When I go to a country, I do not ask if that country has good or bad laws, since good and bad laws every country has, but if they enforce their laws."

Some people use the term conscious to designate failures committed by people who intend to cause harm. We consider most appropriate using the term to designate

Human Failures

deviations from standard procedures. Whoever commits the fault does not want to cause damage and loss. In most cases, they "think it will work" and assume the risk. Those who travel by car without a spare tire know that the procedure is inadequate, but do not aim at the consequences. The term "premeditated" is also inadequate because people react negatively, feeling accused of sabotage. Imagine the reaction of somebody before an analyst who provides the diagnosis: "You committed a premeditated failure." The analysis and prevention will be impaired. Let us reserve this denomination for the purposeful failure when the individual really aims the damages as the sabotage.

Facing the choice between two procedures, the individual (or group) enters in a conflict with himself, and the decision depends on the balance of forces acting in the direction of the standard and of the alternative procedure. Understanding these forces and their intensities is critical for the analysis and treatment of conscious failures. The apparent strength and the real one may be different. Let us analyze an example. A manager installs posters, publishes newsletters, and gives lectures on safety. However, through a nonverbal language, much more emphatic than the written or spoken ones, he transmits his real priorities. If day-to-day activities and concerns reveal that he only values production and costs, the workers "get the message," which reinforces those standards to the detriment of safety, despite posters and speeches.

Often, conscious failures present characteristics of stability. They look like a technical failure. The difference is that in a technical failure the individual does not do correctly because he cannot or does not know, and in a conscious failure, he does not do it because he does not want to. Let us see some examples.

1. An individual opens a bottle by striking the lid against the table's edge, breaking the bottle and cutting off the hand. He had not found the opener.
2. A driver advances the red light at an intersection and causes an accident. He was late for an appointment.
3. A person commits suicide. In this case, the objective is the damage.
4. A house cleaner breaks a glass while washing dishes. If the frequency is one per year, it is likely she fails by carelessness. If it is once a week, it is a technical failure (lack of skills, inadequate procedure, or precarious facilities), or it is a conscious failure (she works hurriedly to finish work earlier or deliberately breaks down a glass for some reasons).
5. A worker removes the safety glasses while working and a spark strikes his eye. Many people who do not use safety goggles had visited the workshop (intruders).
6. An operator performs a task in half the standard time and damages a piece of equipment. His supervisor had told him that he could do it in a shorter time to speed the service (command).

By analogy with conscious failures, let us call failure C, which occurs when equipment can perform right its function but does wrong because of improper adjustment of control systems. This type of failure occurs with flow control valves operating under the command of controllers.

9.4.2 Conscious Failure Mechanism

Behavior is the action that man practices when interacting with the world. A person throws a cigarette on the ground, crosses a street, and uses a safety helmet. Behavior is an observable act that we can record and even measure.

Let us analyze the following smoker's behaviors. At the train station, he throws a cigarette on the floor. At the subway station, he puts the cigarette in an ashtray. Externally, we see the difference between behaviors. How about internally? Something changed when the individual passed from one place to another. However, we have no reason to believe that he has undergone any transformation since no one changes so quickly. Understanding what has occurred depends, basically, on three concepts: attitude, posture, and behavior's consequences.

9.4.3 Attitude

Attitude is an internal predisposition, resulting from several factors, such as values, beliefs, fondness, knowledge, information, and the behavior itself. Receiving proper education, growing in an environment that values cleanliness and discipline, the individual develops a natural predisposition to place cigarettes in ashtrays and trash in dumps. Since his motivation comes from inside, he adopts the same behavior in the crowd or when alone.

Attitude evaluation requires special techniques and yet the result may have considerable deviations from reality. This is because the evaluation requires surveying an individual's mind, and it is impossible to do this fully and objectively. They say the human mind is unfathomable. People can lie, say half-truths, disguise, and deceive the appraiser. Moreover, subjective factors intervene, for they will be observing other's interior through their paradigms. Seeing an individual using the shirt untucked (unusual behavior), a colleague concluded that he was in a relaxed state. However, his pants' back pocket was torn and the shirt was just covering it. Therefore, caution is necessary when making inferences about attitudes through the observed behavior.

Values strongly influence attitudes. Value is something important to the individual, and he tends to act defending his values. The movie "The Bridge on River Kwai" shows a notable case. Japanese troops were forcing British prisoners to build a bridge that would transport Japanese army's supplies. At the beginning, they did not cooperate. The British soldiers frequently sabotaged the work. However, when the British commander decides to show the British competence (value!) to the world, constructing the unwanted bridge, the soldiers change their attitude, and consequently their behavior.

Modifying values is difficult because they develop from individual's childhood. However, the organization's values can change because they are the values that the leaders demonstrate. If people's life and health are a manager's value (not just his own life and health), his attitude and consequently his behavior will promote employees' safety.

A multi-fatal accident that occurred years ago in Brazil shows how values determine individual and organizational behavior. "An armored car crashed from a barge

Human Failures

with four occupants. Firefighters practically discard finding someone alive (O Estado de São Paulo, Tuesday, December 23 of 1997)." The armored vehicle fell from the ferry that was carrying it. Four security men were inside. At 26 m deep, the pressure is 2.6 kgf/cm^2, and the resulting force against a one-square-meter door is 26 tons. No way to open the door.

Keeping people inside the armored car crossing the channel reveals the transportation company's values. Although some ignorance factor was present (no risk analysis), the accident showed that money, not human life, was the value.

Fondness is another powerful factor in the development of attitude. Fondness is a psychic phenomenon that manifests as feelings and emotions, accompanied by satisfaction, pleasure, love, friendship, and care. It may be present in person–person and person–object relations.

If the organization's components have no fondness toward safety function, or worse, if a negative feeling exists, it will be difficult to develop favorable attitudes toward accident prevention and emergency control. The fondness responds better than values to intervention actions, but they are much more refractory than beliefs.

Associating an object with characters or people who already have people's fondness is an effective strategy. Mickey, Minnie, Woodpecker, babies, and children have people's fondness.

People develop fondness for what they create. This fact suggests a strong strategy of lasting effect that to promote the participation. One can create fondness for an object by promoting participation in its creation and development. The River Kwai Bridge's construction developed the British commander's fondness toward the bridge, and when a British command attempted destroying it, he, surprisingly, tried to avoid it.

A third factor in generating attitudes are the beliefs. What we consider right or wrong, true or false, convenient or inconvenient, useful or useless promotes attitude. If people believe that "accidents are inevitable," they will not develop favorable attitudes to behave safely.

Beliefs are refractory to changes, but it is easier modifying them than values and fondness. Information, authorities' opinion, and leaders' posture are efficient instruments for modification of beliefs.

Information on reduction of accidents due to implementation of standard operating procedures undermines the foundations of the belief "Accidents are inevitable."

Authorities are persons who accomplished great deeds. Their opinions impact people's beliefs. People listen and accept Amyr Klink's[2] advice regarding safety planning. However, the same advice goes unnoticed when company's safety technician issues them.

Individuals tend to consider the opinions of their leaders, especially when they are also natural leaders.

A fourth factor, heroes and idols, influences people's attitudes. People tend to approve their heroes' behavior or even imitate them. Therefore, we need to know people's heroes to plan modification of attitudes. The stories people tell and mainly the way they tell reveal who their heroes and idols are. If people admire an employee who defied safety rules, their idols influence negatively their attitudes.

The fifth notable factor in developing attitudes is behavior itself. When an individual spends some time behaving in a certain way, he develops attitude consistent with the behavior. Therefore, attitude generates behavior, and behavior generates attitude.

Beliefs, values, fondness, and idols are the control variables, and attitude is the controlled variable. The manipulated variables are information, knowledge, authorities' opinion, affections, participation, and leaders' examples.

Changing attitudes requires a time-consuming process, and in many cases, short- or medium-term changes are impracticable. To obtain fast results, the control must implement behavior's consequences. The behavior itself will change attitudes, refeeding the control system.

9.4.4 Posture

Posture is the exteriorization of attitude. We may say that the attitude is inside the house, the posture is at the door, and the behavior is in the street. No one can see the attitude, even if the doors and windows are open. Posture is visible. When a person approves or disapproves a behavior, we say that he or she adopts a posture. However, posture does not guarantee that the attitude is consistent with it or that the behavior will be. If an individual adopts a posture under pressure, he may behave incoherently with it at the time of action.

Leaders' posture is a critical factor for safety, but much more important is their behavior. Managers' posture and behavior promote team's attitudes and behavior.

9.4.5 Behavior's Consequences

At the subway station, the man who put the cigarette in the ashtray could want to throw it on the floor. Why did not he does it? Behavior's consequences are the reason. People would throw disapproving looks, or the police would punish him. Therefore, besides attitudes, behavior's consequences control the behavior itself. Moreover, consequences are faster to produce results. However, associating attitudes with behavior's consequences produces better and lasting results.

Behavior's control through the consequences involves antecedents, behavior, and consequences. An antecedent is something that precedes and triggers the behavior; behavior is the observable act, and a consequence is something that stems directly from behavior.

When a doorbell rings (antecedent), someone goes to the door (behavior) and talks to whoever touched (consequence). What controls behavior? First answer could be the bell ringing. Of course, the bell ring has something to do with behavior. If the doorbell did not ring, no one would go to the door. However, suppose the doorbell rings, someone goes to the door and finds no one. Suppose this happens a few times in a short time. After a certain ringing number, the bell will not motivate anyone to open the door. In fact, the consequences, not the antecedents, control behavior. Antecedents are effective in triggering behavior as long as they predict the consequences. The power of the antecedent in triggering the behavior stems directly from the historic of its ability to lead to the consequences.

Any behavior has more than a single consequence. Generally, the number is high, but only a few are of interest. In addition, different consequences may command the

Human Failures

behavior in different directions. Cases of greater interest are those that admit antagonistic behaviors, one is safe while the other is insecure; one is lawful, while the other is illegal. The dispute between their consequences demands the individual's decision, and he will adopt the behavior whose consequences have greater control power.

Time, consistency, and significance affect the influence of behavior's consequences. Time's influence varies as the consequence occurs immediately or later after the behavior; consistency, as the consequence is certain or uncertain; and significance, as the consequence is positive or negative.

The consequences of greater power are immediate, certain, and positive. Negatives and positives have the same strength, but the side effects are different. The negative ones reduce engagement and morale and have adverse effects on attitude. The positive, by contrast, increase engagement and morale and improve attitude. The ideal composition is a mixture of positive and negative consequences, strongly focused on the positive side.

Some cases seem not to find an explanation in the consequences model, but it is still applicable if we consider the concept of virtual consequence. Virtual consequence exists in the individual's mind; it is not real but has great power to control behavior. Advertising associates smoking with virility, freedom, and elegance. An individual who associates these qualities with smoking experiences positive and immediate consequences lighting a cigarette and aspiring the smoke. Positive, immediate, and certain, it has great control power. With the habit arises a real consequence, the chemical dependency.

Smoking's negative consequence, the cancer, has little power to control the behavior. Cancer is not a hundred-per-cent-probability event, or follows immediately the behavior, but years later. Therefore, it is remote, and people smoke. However, no advertising would succeed in promoting the unsafe behavior—putting hands on a 440-volt energized wire—even associating positive consequences, because the negative effect is immediate and certain.

Let us analyze an example, addressing the emery activity. The standard procedure requires safety glasses, and warning signs indicate that its use is mandatory. The safe behavior is to wear safety glasses, but it has a negative consequence, the discomfort.

The alternative procedure is to work without glasses, an insecure behavior, no protection. It also includes a negative consequence, vision loss, but has a positive one, the comfort when removing the glasses. Apparently, the unsafe behavior's negative consequence has greater control power, as comfort does not compensate the loss of vision. So why workers adopt unsafe behaviors and do not use safety goggles? The answer lies in consequence's attributes. Vision loss is negative and discourages unsafe behavior, but it is uncertain and remote. In turn, the comfort is immediate and certain. Loss of vision is a probabilistic event. It may or may not occur. Moreover, it is not immediate and may occur after a certain time working without protection. In addition, it is something negative, and workers tend to dispel that possibility. Therefore, the consequence is remote, uncertain, and negative. In turn, comfort is a certain and immediate consequence of greater control power.

The warning plate is an antecedent that aims to stimulate the safe behavior. This antecedent has no power since it has no connection to any consequence. Best would be to remove the plate. Unsafe behavior's driving forces win, and a worker will certainly swell the disabling accidents statistics.

The intervention must include, among other control measures:

1. Providing comfortable safety glasses to reduce unsafe behavior's positive consequence.
2. Training supervisors so that they can understand their role in promoting safe behaviors. Supervisors must impose consequences certain, immediate, and negative to the unsafe behavior (reprimand, punishment), and consequences certain, immediate, and positive to the standard procedure (praise, promotion).
3. Implementing safe behavior indicators.
4. Promoting changes of attitude, focusing on values, beliefs, and fondness.

When the standard procedure is incorporated into the organizational culture, offenders suffer certain and immediate consequences, the disapproval of colleagues. Cultural strength is powerful because it is everywhere and acts all the time.

Conscious failure analysis through the behavior's consequences model compares conflicting behaviors' consequences and antecedents. Control measures must add certain, immediate, and positive consequences to safe behaviors; certain, immediate, and negative consequences to unsafe behaviors; and antecedents that have an effective connection with the consequences.

9.4.6 Safe Behavior's Driving Force

Let us call F_s the force that drives a person toward safe behavior. The higher the ratio (product of the impulsive factors)/(product of the restrictive factors), the stronger the driving force. Here are some considerations about those factors.

1. Behavior's negative consequences

    The price is a negative consequence. Each behavior has an associated price. In addition to monetary amounts, it includes physical, mental, or emotional effort. Inadequate price evaluation or lack of knowledge allows finding cheaper alternatives that contribute to the adoption of insecure behaviors.

    The risk is a negative consequence. This consequence is uncertain and remote since risk is a potential loss or damage, not a current reality, but a virtual one. Therefore, whoever evaluates risks does not have the reality, but a mental representation. Moreover, even for people who have suffered damage, the reality is only present at the time of the accident. Over time, the mental representation changes. When exacerbated, it creates neurotic behaviors; attenuated, it loses the control power.

    Imagination or suggestion can create a favorable mental representation. Some "mind control" methods create programmed representations of reality. The representation, a perceived risk, commands behavior.

Human Failures

2. Behavior's positive consequences

Positive consequences produce satisfaction. Each behavior has an associated satisfaction. The evaluation, or rather inadequate perception, of the behavior's quality promotes the adoption of unsafe behaviors. A worker may experience satisfaction by not wearing a seat belt and perceiving "admiration" from colleagues who consider him brave and bold.

One cannot dissociate the conscious failures from technical factors since human behavior results from risk, price, and quality assessment. Let us develop an expression to indicate the direction each factor drives the behavior. It is a tool to facilitate understanding the involved phenomena, not a formula for calculations.

The following equation expresses the intensity of the driving force:

$$F_s = (A_s/A_i)_* (S_s/S_u)_* (C_{S+}/C_{S-})_* (C_{u-}/C_{u+})_* (R_u/R_s) \quad (9.1)$$

where
C_{s+} = safe behavior's positive consequences
C_{s-} = safe behavior's negative consequences
C_{u+} = unsafe behavior's positive consequences
C_{u-} = unsafe behavior's negative consequences
R_s = safe behavior's perceived risk
R_u = unsafe behavior's perceived risk
A_i = unsafe attitude, predisposition to unsafe behavior
A_s = safe attitude, predisposition to safe behavior
S_u = social rule approving the unsafe behavior
S_s = social rule approving the safe behavior
F_s = safe behavior's driving force

Factors C_{s+}, A_s, and S_s produce satisfaction. Therefore, we can replace the product $[C_{s+}] [A_s] [S_s]$ by Q_s, safe behavior's quality. R_s and C_{s-} are components of the safe behavior's price. Therefore, we can replace the product $[R_s] [C_{s-}]$ by P_s, the safe behavior's price. In the same way, we substitute $[A_u] [S_u] [C_{u+}]$ by Q_u and $[R_u] [C_{u-}]$ by P_u. Equation 9.1 takes the form of the following equation:

$$F_s = (Q_s/P_s)/(Q_u/P_u) \quad (9.2)$$

However, $Q_s/P_s = V_s$, safe behavior's absolute value and $Q_u/P_u = V_u$, unsafe behavior's absolute value, resulting the following equation:

$$F_s = (V_s/V_u) \quad (9.3)$$

Thus, the safe behavior's driving force is its relative value when the unsafe behavior is the reference. An individual adopts the behavior of greater relative value. The customer (the person who adopts the behavior) will buy the safe behavior if its relative value increases. If V_u = zero, the safe behavior driving force goes to infinity (becomes very high). To make V_u = zero, we can make the quality of the unsafe behavior very

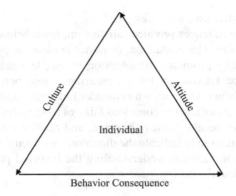

FIGURE 9.1 Interactions—culture, attitude, and behavior.

low, or its price very high. V_u may be null if the unsafe behavior does not exist. In this case, the safe behavior's relative value goes to infinity for "lack of competition." We can approach this mathematical reality by reducing as much as possible the unsafe behavior's probability, designing systems that make it impracticable.

Figure 9.1 represents interactions between attitude, behavior consequence, and culture. Acting on the individual, they generate the behavior's driving force. However, when more than one behavior is possible, the individual will choose one of them, and the choice depends on the balance of the driving forces. The individual adopts the behavior of greater driving force. Therefore, the behavior's driving force study must precede the analysis of the process leading to the unsafe behavior.

9.4.7 Behavior Control

Since human actions have a significant contribution to a great number of accidents, the risk control depends largely on human behavior control. Behavior control requires power. Power comes from three components: authority, accountability, and resources. One must consider the 12 resources analyzed in Chapter 2.

We can control the behavior through the factors that command it: attitudes, postures, and behavior's consequences. However, human nature is complex and diverse. Therefore, the result of the intervention is not a certain event, although we can achieve it with high probability. Persistent deviations require special treatments.

A conscious failure results from a decision-making choosing the unsafe behavior when a safe one is available. Adopting a standard procedure is the safe behavior since it has a lower associated risk. Choosing the alternative procedure is an unsafe behavior since it has a higher associated risk. Understanding the mechanisms that control the conscious failures is to understand the mechanisms that control human behavior.

Standard is a benchmark for evaluation. A deviation is the difference between the observed value and a standard. Process control takes the measured variable to the desired value defined as standard. Similarly, behavior control takes the observed behavior to a desired one defined as standard. The behavior that has a tolerable and

Human Failures

as low as practicable associated risk must be the standard because no behavior has a null risk. Control actions must follow the holistic paradigm, approaching behavior as a multifaceted phenomenon, resulting from complex interactions between physical, biological, psychological, cultural, and social factors. Like any control system, behavior control must have standards, sensors, and controllers. Although cameras could act as sensors, the best sensor is the human eye, especially the supervisors and managers.

A good sensor must detect accidents and risk factors. A large number of behavior deviations occur before an accident. The analysis of abnormal occurrences must consider an underreporting percentage caused by people's disbelief in the system effectiveness or its goals.

A controller is the device that receives information from sensors, compares the measured values with the standard, and introduces intervention actions as a function of the deviation. Actions depend on the reaction model of the controlled system and the controller's own model. As a rule, one must add consequences certain, positive, and immediate to the desired behavior, and consequences certain, immediate, and negative to the unwanted behavior. The greatest control power lies in the immediate character. A consequence certain, but remote, does not have the same power. Therefore, management must show immediately the willingness of not tolerating violations of safety standards.

To determine intervention actions, it is necessary to know how the social system reacts to interventions. For this, one must know the social system in all its dimensions: physical, biological, cultural, social, and economic. Values, policies, guidelines, programs, projects, and strategies are control variables. Moreover, to establish the intensity of the control actions and the application speed, the controller must choose one or a control models composition.

The controller may expect the variable to reach limit values before any intervention. This is the "on–off" control that is not appropriate for behavior. Physical variables recover the initial values after control actions, but behavior generates attitude, and attitude generates behavior what may make the recovery impracticable.

The controller may act whenever a deviation occurs. This feedback control compensates the system when the disturbance's effects occur. The feedback control may make interventions proportional to the deviation or consider the deviation accumulated in time and the speed with which the deviation progresses.

Small changes may generate the "boiled frog" situation. They say that a frog thrown into boiling water reacts, jumping and escaping. However, placed in cold water, it will die without reacting, if one heats the water slowly. The man reacts to changes of a certain magnitude, not to minor changes. Therefore, one must consider the control actions based on accumulated deviation when controlling behavior.

Social systems react slowly to disturbances. Therefore, the control must consider the variation rate of the deviation. If the behavior changes rapidly, some factor has been acting and its effects will continue to grow, requiring fast countermeasures. It may require also an anticipatory control intervening before disturbance effects occur.

Behavior requires a complex multivariable control system because cause-and-effect relationships are probabilistic.

Behavior control requires a cascade action. Increasing training hours (first variable), we can provide more information (second variable), which improves attitude (third variable) one of the variables that control behavior (final variable).

A controller can manipulate several variables to perform an intervention. Choosing a variable requires a criterion, which involves a parameter and a rule. The parameter can be the variable's gain, and the rule, acting on the highest gain variable. K (gain) is the ratio: (controlled variable's variation)/(control variable's variation).

Investing in a certain variable, the gain tends to decrease until it reaches the saturation point when no additional variation occurs or the associated cost is very high. Each variable has an optimal point, from which it is better to invest in another.

9.4.8 Leadership and Behavior

How are leaders behaving? The answer to this question explains most of the occurrences involving unsafe behavior. By action or omission, leaders can give negative examples. Formal leaders play a vital role in controlling behavior. As the ones led by them tend to mimic the leader's behavior, the example is one of the most effective instruments of control.

Analysis of leadership's posture and behavior requires identifying their values, beliefs, and feelings. For this, we must consider nonverbal language, often stronger than the verbal one. *Who silence is consent* applies here. If a leader is silent when observing inappropriate behavior, the ones led by him understand that he approves or tolerates that behavior.

9.4.9 Cultural Change

The most immediate and certain behavior's consequence comes from culture. Organizational culture has a strong influence on behavior because an individual needs his group's acceptance, and culture exerts a powerful force on him.

Behavior's consequences model explains the influence of the organizational culture since countering cultural rules brings certain, immediate, and negative consequences. In the medium- and long term, behavior control must aim at cultural change, the change of the collective behavior.

Every change requires a new paradigm. It is necessary to revolutionize the way of thinking, facing reality, and acting. As a result, attitudes, postures, and behaviors change. However, not everyone follows that order in the change process. Initially, they adopt the new behavior to avoid negative consequences as happened with vehicles' seat belts in some countries. People started using the belt to avoid fines. Moreover, some drivers used fake belts to "trick" the police. However, after some time, advantages of the belt became so evident that they promoted great attitude and posture change. When this happens, people adopt the behavior even in the absence of negative consequences. Changing process is complete and is not difficult to maintain the new situation.

When a community, society, or organization changes their attitude and behavior, a cultural change occurs.

Human Failures

9.4.10 Conscious Failure Treatment

Increasing the relative value of the safe behavior is the strategy of conscious failures' treatment. For this, it is fundamental:

1. *Establishing standards and objectives.* To avoid demoralization, they must be realistic. Conscious failure consists of taking a risk greater than the tolerated one. A well-defined pattern helps the failure characterization. If the pattern changes, usual practices may become failures.
2. *Establishing priorities to reduce conflicts.* Seeking consistency between discourse and practice, avoiding saying one thing and practicing another.
3. *Establishing responsibilities and accountabilities.* Who is responsible for the task? If no one has to answer for the correct procedure, people will probably modify it.
4. Designing equipment and facilities that can eliminate the possibilities of unsafe behavior.
5. Creating immediate, certain, and positive consequences for safe behaviors, and immediate, certain, and negative consequences for unsafe behaviors.

9.5 COMPOUND FAILURE

Failure boundaries are not lines, and in practice, all failures result from a composition of the basic types. Unfolding failures in their components helps to define the preventive actions according to their relative contributions. If we say that a failure is of a certain type, we are referring to the predominant factor. Analyzing conscious failures, we may find that an unsafe alternative procedure resulted from poor knowledge. An individual thought that he had the knowledge and control, and that predicted and evaluated all the consequences. In fact, he knew that was violating the standard, but ignored the accident's probability. Technical failure! Therefore, conscious failures result from technical and conscious factors. Carelessness failure may reveal a technical face if we find that the individual was likely to fail under the working conditions. When we identify risk situations that may cause accidents if someone commits oversights and we do not correct them, an accident would result from carelessness and a conscious component. Moreover, the conscious component may contain ignorance about consequences and occurrence probabilities, a technical component.

9.6 FAILURE PROMOTORS

Primary, secondary, command, and intruder agents promote human and equipment failures. Generally, promoters act in combination. In some cases, one of them can be so prevalent that the others become negligible. It may occur that all they contribute significantly, and with synergistic effect. Figure 9.2 represents graphically the joint action of these factors.

9.6.1 Primary Agent

Primary failure is the one that occurs in the environment and under load for which the component has qualification. Example: A pressure vessel ruptures under a pressure

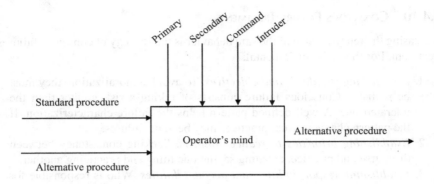

FIGURE 9.2 Failure promoters.

lower than its design pressure. Generally, design, fabrication or construction, aging, or maintenance causes primary failures.

Similarly, a primary failure occurs when the individual is or becomes incapable of performing the design function under design conditions. Design function is the one that the individual can execute, as long as the system and the environment remain in the normal design conditions. Failure comes from the individual.

Treating primary failure, preventive action must address the individual who committed it. We have reason to believe that another person placed in the same situation would not commit the failure. As a rule, one should not attribute the cause to the primary agent without analyzing the contribution of other agents. Let us analyze some typical cases.

1. A worker does not separate properly green and red objects. He is color-blind. The primary agent promotes the technical failure.
2. A worker forgets to perform a task. He is separated from his wife. Primary agent promotes the carelessness failure.
3. The worker climbs on the scaffold without using a safety belt. He thinks he has great skill and can dispense the belt. Primary agent promotes the conscious failure.

9.6.2 SECONDARY AGENT

Secondary failure is that occurring in the environment and under load for which the component has no qualification. A pressure vessel ruptures under a pressure higher than design pressure. As the name implies, the primary cause does not stem from the pressure vessel, but from the excessive load or the adverse environment. Similarly, the human being commits secondary failures when working under adverse conditions, such as inadequate layout, heat, noise, over time, and extreme psychological stress. Corrective action must aim those conditions, not the person who commits the failure. Other persons placed under the same conditions would fail. Let us analyze some typical cases.

2. A pedestrian crosses a highway, and a car kills him. The only footbridge is 5 km away. A secondary agent promoted the conscious failure.
3. An operator forgets to perform routine checks and does not detect the high level in a storage tank. The tank overflows. This operator has been doing the same job for several years. A secondary agent, the work routine, promoted the carelessness failure because people who perform routine work become careless after some time.

9.6.3 Command Agent

A component commits a command failure when acts incorrectly obeying orders of another system's component. The controller sends a wrong signal, and the control valve blocks the inert gas flow. Wrong actions resulting from the project are command failures. A relief valve discharges toxic gas into the operating room's air intake. The valve commits failure, but the project commanded its action. Similarly, a man commits failures under the command of supervisors, written procedures, and organizational culture. Warning signs exert command over people. In general, command failures result from wrong information.

People act under the influence of several commands. In addition, some commands are not explicit. Through nonverbal language, managers give nonverbal commands, and the day-to-day small actions are stronger than the official position. If, during the visit to the industrial plant, the top management only makes questions about production, ignoring the safety, they pass their true priorities to the staff. Omission prioritizes production at the expense of safety.

A notable case comes from literature: "It happened years ago when I was working in the operation and a new manager started working. After a few months, someone said that employees considered him less interested in safety than his predecessor. 'What did I say,' he asked, 'to make that impression? Please convince everyone that this is wrong.'" It was not what he said that created the impression, but what he had not said. Let us analyze some command failures.

1. A driver approaches the intersection where a police officer is guiding the passage of vehicles. Despite the green light for vehicles, the officer signals *stop*. The driver confuses himself, crosses and collides with another vehicle. Dual command, officer, and light promoted a carelessness failure.
2. An operator receives several orders and counter-orders to put pump A into operation. Finally, he receives instruction to place pump B. When performing the operation, he confuses and places pump A. The supervisor promoted the carelessness failure.
3. An operator puts $10\,m^3$ of water into a tank when he should have put $5\,m^3$. The incorrect procedure indicated $10\,m^3$. The procedure promoted the technical failure.

An electrician does not use the warning label when performing maintenance on a motor. He joined a work group that does not use warning labels. Organizational culture promoted the conscious failure.

A welder does not wear protective goggles. His supervisor has seen other welders without safety glasses and kept silent. The supervisor promoted the conscious failure. Omitting himself, he gave a nonverbal command.

9.6.4 Intruder Agent

Man can be an intruder causing system failures, disabling protection devices, and changing safety alarm, interlock set point, or valve position. By extension, let us consider as intruder anything foreign to the normal operation and environment such as personnel performing maintenance, inspection, and visits; events of nature, such as floods, storms, and windfall; political events, news and rumors. Let us analyze some intruder cases.

1. An operator eliminates steps of the procedure and causes an accident. A colleague from another area, claiming to be knowledgeable, had made suggestions. The colleague promoted conscious failure.
2. An operator makes mistakes when controlling a panel during an emergency. Upon hearing of the emergency, several people had arrived to help with suggestions, stunning the operator. Stranger to the operating system, including area managers, promoted the carelessness failure.

9.7 COMMON CAUSE FAILURE

Redundant elements increase systems reliability. However, it cannot increase indefinitely due to common cause failures (CCFs). Simultaneous action of failure promoters on redundant elements creates CCFs, reducing or even nullifying the desired reliability increase. The CCFs determine the reliability of a high-reliability system.

What makes systems vulnerable to this type of failure are the common susceptibility and common location. Five parallel relays aim to increase an electrical system reliability. However, the electrician calibrates all relays incorrectly. Therefore, a single cause brings simultaneous failure of all redundant elements. CCFs determine some procedures. In a republican country, the president, vice president, and the deputy chamber's president never travel together in the same airplane.

Reliability studies consider the CCFs. In turn, qualitative checklists can detect the factors that cause them. A simple exercise can greatly help increase the ability to detect CCF promoters. Its purpose is to sharpen the ability of observation finding what is common among various sequences of letters. What do the letter sequences have in common?

1. A A B B C C	2. B G J O P	3. K L M N T V	4. D Q U C G
5. A E F H I K	6. A C E G I K	7. K K L L M M	8. L N P R T

The answers can be one and seven repeat letters in sequence in the alphabet; three, five, and seven have rectilinear line letters; two and four have curved lines letters; six and eight result from eliminating the letters immediately following each letter in the alphabet sequence. However, no matter how hard we try, we will observe a limited number of a set of infinite possibilities. This is because we have imposed restrictions not established in the statement of the exercise. Removing the constraints, we can identify that all sequences have the same color and are on the same paper, same book, same room, same city, the planet Earth, etc.

This exercise prepares to observe common causes of failures of redundant elements. Consider two centrifugal pumps of the same industrial plant, one parallel redundancy of another. What is common? Both have electric motors, receive electricity from the same substation, are at the area, have the same supplier and manufacturer, pump the same fluid, and are maintained by the same technician. We can thus identify common causes and establish measures to reduce the likelihood of simultaneous failure. One of the pumps could have a steam turbine and/or a different mechanic technician could perform the maintenance.

NOTES

1. Boilover is a phenomenon during burning of certain oils in an open-top tank when, after a long period of quiescent burning, there is a sudden increase in fire intensity associated with expulsion of burning oil from the tank.
2. Amyr Klink (born on September 25, 1955, in São Paulo, Brazil) is an explorer, sailor, and writer. Amyr Klink was the first person to row across the South Atlantic, leaving from Lüderitz, Namibia, on June 10, 1984 and arriving 100 days later in Salvador, Brazil, on September 18, 1984.

BIBLIOGRAPHY

1. Kletz, Trevor A. 1985. *An Engineer's View of Human Error*. England: Institution of Chemical Engineers.

10 Safety Concepts

Theories are not false or true; they are useful or useless[1]

Sanders have devices called machine guards (protection). They block the pieces of disc in case of breakage and provide mechanical isolation, preventing the operator from touching the disc. Therefore, those devices do not protect the operator. Protection is a device such as gloves of steel mesh that protect the operator. We are not claiming that those concepts are wrong. However, concepts must form a coherent and integrated set as requires a holistic approach.

Let us present an example using the safety concepts, which we indicate in italics for reference. We recommend that the reader return to this example when completing the study of concepts.

Hydrogen sulfide is a gas whose chemical formula is H_2S. Depending on the concentration, the inhalation causes from discomfort to instant death. Its rotten egg odor is a *natural alarm*, but it is an insidious gas since it anesthetizes the sense of smell, and people no longer can smell it.

Let us consider a pressure vessel containing hydrogen sulfide as an *object of study*. This pressure vessel is a *source* of *aggressive agent*. The *containment system* (pressure vessel) prevents hydrogen sulfide from reaching *targets* (people). *Rupture agents*, such as corrosive atmosphere and the own hydrogen sulfide, act on the *containment*. A *restoration system*, such as inspection and maintenance, neutralizes those *rupture agents*. If the *restoration* fails, leaks occur and the *aggressive agent* creates an *aggressive action field*.

The leak is a *hazardous event* that initiates an *emergency* creating a gas cloud, an *impactful agent*, and also an *aggressive agent*. The *impact effects* on *targets* depend on agent's *aggressive capacity*, a function of the amount of gas and level of *aggressive energy* (hydrogen sulfide concentration in the air).

Exposure of unprotected *target* is a *damaging event*. The cloud injects the *harmful agent* (H_2S) into the *targets*. *Protection systems* (gas masks) could have prevented this event. To avoid people entering in the *aggressive field*, the *safety group* installs an *isolation system* (barriers, ropes) and warning signs (*artificial alarms*). *Containment*, *isolation*, and *protection* have different functions. Containment prevents the release of aggressive agent; isolation, entry of targets into the aggressive field; and protection, the agent penetration into the target.

Some people in the *aggressive field* smell the rotten egg odor (*natural alarm*), but the gas neutralizes their *detection system* (sense of smell). Due to the respiratory tract *vulnerability* to the gas cloud, the *harmful agent* penetrates the *targets*. As the blood has a high *susceptibility* to the *harmful agent* and the *defense system* (metabolism) has no *power* to eliminate it, *damage* occurs.

The emergency *control system* triggers audible *alarms* (*evacuation system*) and some unaffected people leave the area, while the *rescue group* removes others and hand them over to the *medical assistance group*. After medical care, they refer the *victims* to the hospital (*recovery system*). One person dies (*damage*), and the family members suffer a *loss*. The *combat system*, using water mist, disperses the gas. The *abnormal occurrence analysis* recommends *risk control* measures aimed at improving the *reliability* of the pressure vessel. Reviewing the existing *risk assessment*, management expands the *hazardous area*.

10.1 OBJECT OF STUDY

Everything related to events that harm people, environment, or property, causing physical, psychological, or moral damages, properties or environmental degradation, and various types of losses can be the object of safety study. We study an object to know or transform it.

Any being, animate or inanimate, natural or artificial, concrete or abstract, real or fictitious, can be candidate for safety study. Beliefs, values, and team morale are abstract objects. Ghosts or zombies, among others, are fictitious objects that some people fear or use as excuse to avoid working in some areas at night.

Organizations, systems (mechanical, electrical), activities (welding, carpentry), actions and relationships between beings, events or sets of events, past, future, real or virtual, aggregates, or systems can be the object of study. The approach can be reductionist or systemic.

Let us call anatomy the aspects related to the object's structure and form. The structure is the arrangement; form is the configuration or particular aspect.

Physiology is the set of aspects related to functions, processes, and activities. A function is everything the object or its components do (paint, transmit, cut, weld). A process is a set of organized functions. Activity is any work with repetitive characteristics (carpentry, welding, surveillance).

10.2 AGGREGATE

Aggregate's properties result from parts' properties. Adding components' properties, one obtains the whole's properties. The aggregate is the sum of its parts. No proper interconnection exists between the components, no matter how one organizes the parts. The essence of the aggregate does not change adding or removing some parts, such as a set of bricks and a sand pile.

10.3 SYSTEM

System is an entity composed of interrelated and interdependent elements that transform stimuli received from outside with a defined purpose. Components' interactions create new features, absent in isolated or grouped components. Withdrawing components alters or destroys system's essence. Relationships are more important than the parts.

Inputs, processes, products, standards, feedback, sensors, and regulatory function (controller) are the elements of a system.

Inputs come from the environment. They can be active or passive: passive inputs (raw materials) receive the action of the process to become products, and active inputs (energy, equipment) are essential to perform functions.

Safety Concepts

A process is a set of organized functions, a functional structure with defined relations. The process may be unitary, consisting of a single function, or multifunctional, consisting of two or more functions. Each component has a basic or main function and a set of complementary and auxiliary ones. Functions and their relations are not sufficient to characterize the process. Process variables such as temperature, pressure, fluid flow, chemical properties, organizational climate, and adherence determine the conditions under which the system performs functions.

A product is the result of the transformation of inputs. It can be equipment, material, energy, installation, or risk. Intermediary functions generate intermediate products, and the whole process generates the final product. Processes may produce unwanted products along with the desired ones. The unwanted products can be innocuous or harmful. Waste and pollutants are harmful products. Service is a human action that does not assume the form of material goods.

The evaluation of system performance requires benchmarks, the standards for productivity, quality, safety, and environmental preservation. The sensors measure the attributes of the system and inform the controller.

Feedback is an output information that refeeds the process. It causes changes of inputs, processes, products, and patterns. Negative feedback cycle reacts to cancel the changes. Positive feedback loop amplifies any system disturbance. Feedback cycle reaction time varies depending on system type. They can be short (pain reflex), medium (temperature control with thermostat), or long (social systems).

A regulatory function governs the internal and external interactions, controlling the process. The controller performs the regulatory function, which can be simple in some cases and complex in others. Receiving stimuli from outside, the controller defines functions, their direction, application point, and performance, and issues commands. Decentralizing and assigning decisions to some components, the controller delegates some decisions.

In continuous operating systems, man has little interference. System structure incorporates most of the regulatory function. Equipment arrangements and the connections define much of the process. However, man modifies the regulatory function by changing variables' set points. In batch processes, and in the shutdowns and start-ups of continuous processes, the procedures contain the regulatory function. Executing the procedures, man is a component; creating them, he is a formulator.

Systems can be mechanical (thermostat temperature control), human–mechanical (man–car), biological (body temperature control), ecological (predator–prey system), and social (company–client). Nature and man build mechanical systems; living systems develop. In safety studies, the most important living systems are the organizations.

10.4 NORMAL STATE

The characterization of the normal state of an object requires a set of variables. Each variable has an expected value or standard and a variance. If the values assumed by a variable are within a certain interval containing the expected value, the state is normal. Otherwise, it is abnormal, and the object has an abnormality. The extent of the normal interval depends on the variance and the observer's choice. A deviation is the difference between the observed state and the expected value. Deviations much smaller than the normal interval are negligible.

The expected value or standard is a subjective parameter. Observer's paradigms, i.e., perceptions, expectations, values, and idealized patterns define the normal state. Normality range is a choice of the observer and depends on the errors he risks to take and the costs he accepts to pay. Therefore, the normality parameters vary with time, place, culture, and needs. Cultural elements reveal the relativity of the normality concept. Some indigenous tribes deform their lips using wooden discs. The deformations, equivalent to serious accidents lesions, are normal in those cultures.

A complete characterization of the normal state requires anatomical and physiological analysis. The object may have anatomical normality and physiological abnormality. This is because the available methods and equipment fail to detect anatomical abnormalities, or the physiological abnormality arises from inappropriate interaction between components.

When the object is a product of man's action and its specification meets needs and interests, the normal state is a conformance and the object is in compliance or conformity. Otherwise, it is a nonconformance or nonconformity.

Physiological failures are abnormalities related to the performance of functions. When the device does not perform the function or execute incorrectly, it fails. Moreover, when an abnormality impairs the proper function performance, the device is in a fault state.

Anatomical abnormalities resulting from design, production, or birth are defects. When the abnormality results from use, the cause of the improper functionality is a fault or damage.

The physiological analysis requires observation over time. The variable to consider is the reliability. Device reliability is the probability that it will perform the function for a given time under given conditions. If we put 1,000 motors to operate for 10,000 h with design load, ambient temperature 45°C, atmosphere free from dust, and one of them fails, we say that the reliability of the set of motors is 0.999.

Deviations from normal state result from wearing, or action of aggressive agents. Aging raises failure rate. This condition ends the object's life cycle, which may also terminate due to the emergence of objects of superior performance.

10.5 ABNORMAL STATE

Damage is the undesirable alteration of the state of the object that results from the action of aggressive agents. Alterations may develop gradually and imperceptibly, or abruptly. It may be anatomical or physiological, reversible or irreversible. When the restoration is feasible, the damage is reversible. Otherwise, it is irreversible. Total reversibility is impracticable, and residual damage remains.

When the alteration is insufficient to affect normality, the damage is negligible, e.g., the state of the skin is a variable that characterizes the normal state of an individual. The mark caused by the smallpox vaccine is a damage, but no one considers having an abnormal skin because of that scar.

Defects arising from manufacturing, birth, design, construction, or assembly are not damages nor the wearing and degradation resulting from the normal use or

Safety Concepts

physiological activity over time. When we speak of "damage caused by the time," it is because we are considering the time as an aggressive agent acting on equipment, installations, and people.

The term defect brings the idea of current state. Damage has the implicit idea that some agent caused the abnormal state. Consider someone trying to return an electric appliance to the seller because it does not work. The same reality, the flawed state, resulted from damage for the seller and from a manufacturing defect for the consumer.

Damages may be personal, patrimonial, or environmental. Personal damages can be anatomical or physiological. Injury is the anatomical damage in men, animals, and vegetables. Psychological and moral damages are more difficult to identify than the physical ones. Moral damage is subjective but considered in court judgments. The physiological or functional alteration may be permanent or temporary and generates signs and symptoms. Sign is objective evidence, such as temperature and pressure, while symptoms are abnormalities that the patient feels, such as headache and tiredness, but a doctor cannot evaluate. Signs and symptoms are changes of state, but not necessarily stem from damages. They may result from reactions to aggressive agents. Environment damages develop environmental degradation.

10.6 LOSS

Loss is the disruption of the owner–object relation. Damage is a change of the object, and loss is a change of the relation possessor–object. When the body suffers damage, the person suffers loss. If a thief steals a car, the owner suffers loss. The electrical energy transformation into heat and dispersion in the environment is a loss.

The loss is repairable when the object is restorable, replaceable, or someone can indemnify the owner. The loss is irreparable when the restoration is impossible or impracticable, or the object is irreplaceable such as body parts, lives, and morale. In such cases, any indemnity has a compensation character.

10.7 ABNORMAL OCCURRENCES

A normal occurrence is a fact or set of facts whose characteristics the observer desires or expects. The characteristics may vary within the limits that determine the normality range.

An abnormal occurrence is any fact or set of facts whose characteristics are outside the normal range. Some abnormal occurrences cause negligible negative consequences while others produce significant damages and/or losses. Accidents, losses, and failures are abnormal occurrences, such as car crash, theft, and fire. Any failure is an abnormal occurrence, although expected with a known frequency, because it is an undesirable event.

When an occurrence is the product of man's action and meets expectations, the normal state is a compliance. Noncompliance or nonconformity is any deviation from the desired state. Therefore, the set of abnormal occurrences contains the nonconformities set. Rainfall much higher than the average is an abnormal occurrence, but not a nonconformity, since man does not determine Nature's behavior. When we acquire four-inch-diameter pipes and receive six-inch pipes, the occurrence is a nonconformity.

10.8 IMPACTFUL AGENT AND TARGET

Let us consider an action involving two entities: an impactful agent performs the action, and the target receives it. Aggression is every action that can cause damage. The impactful agent that can cause damage and loss is an aggressive agent. A target can be a human, a property, or the environment. If the target is a person, let us call it victim. During a fire, the impactful agent (heat) causes injuries to people and damages to components of the environment and property.

An impactful agent may become a target, and the target becomes an impactful agent (reciprocal action). Entering forests, men are impactful agents, but they are targets of mosquitoes. Fish stings injure anglers. An entity can be the target of its own aggressive action (reflexive action). Suicide is the most characteristic example, but it also occurs in every case in which man provides energy to the aggressive agent, such as cuts produced by knives and screwdrivers due to unsafe work at home.

The impact is the agent's action on the target. During the impact, the impactful agent injects matter, energy, or information. Injectable agents may be beneficial, innocuous, or harmful. Impactful agents that inject harmful agents are aggressive agents. Some injectable agents are both beneficial and harmful, such as the medicine that fights a disease, but produces side effects. An impactful agent can simultaneously inject beneficial, harmless, and harmful agents, such as a contaminated needle that injects medicine, water, and a virus.

The effect is any result of impact. Regarding the effects, the impact can be positive, neutral, or negative. The effect can be certain, when the relation impact–effect is deterministic or random when the relation is probabilistic. The risk is a random effect. An impact's effect may be the release of new agents, initiating a domino effect or chain reaction.

Regarding the behavior over time, the effects may be immediate or delayed, transient or permanent. Effects are immediate when manifesting just after the impact. A hailstorm produces an immediate effect on a vegetable garden. Delayed effects manifest after some time from the impact. After contacting a virus, a time elapses before the signs and symptoms of the first sickness appear. Transient effects disappear after some time, the anatomical and/or physiological changes do not remain, and the normal state returns. An oil spill produces a transient degradation of the sea. Permanent effects persist after cessation of disturbances. Vision loss is a permanent effect of the impact of grinding fragments against an eye.

Disturbance is the transition state that follows the impact and can evolve to normal or abnormal states.

1. Normal state, if the variables return to normal range, such as noise exposure causing temporary hearing loss.
2. Adapted normal state, if the variables return to the normal range, but modified to adapt to new conditions, such as people organism moving to high altitude regions or the skin exposed to the sun.

Safety Concepts

3. Evolved abnormal state, when the variables move out of the normal range, but the change is beneficial, such as irrigation in a desert area. This change is an evolution.
4. Degraded abnormal state, if the variables move outside the normal range, and the change is negative, such as noise exposure causing permanent deafness. This change is a degradation.

Assimilative capacity is the ability to absorb impacts. The dose of a harmful agent that a target absorbs without undergoing significant modifications expresses the assimilative capacity. High assimilative capacity reduces the impact effects. Several factors influence assimilative capacities, such as susceptibility, defense system, and mass.

Impact control is the set of actions aimed at avoiding impacts or maintaining their effects below tolerable limits. Environmental impact assessment includes the characterization of the state of the target before the impact, identification of impactful agents and their sources, and impact assessment and definition of control measures. Let us use the safety concepts to analyze a boxing fight. We could do similar analysis by addressing oil spill at sea, fire at a building, or population migration.

Fighter's glove is the aggressive agent, and the target is the opponent. Fighter's arm imparts aggressive energy to the glove. The impact is the shock of the gloves against the opponent's body, and the injectable agent, a mechanical pressure wave, is harmful. The disturbance of the organism is an effect of the impact, and the knockout is a probabilistic effect. Immediate effects are the breath shortness, dizziness, and weakness. They are transient. Loss of strength and fatigue are permanent aftereffects (considering the time of the fight). The effects' intensity depends on fighter's assimilative capacity. Dodging, breathing, and maneuvers to gain time are impact control measures.

10.9 HARMFUL AGENT AND SUSCEPTIBILITY

A harmful agent is the entity made up of matter, energy, or information that causes damages and/or losses when injected into a target. The agent causes harm or promotes the generation of agents that may be more harmful. The harmful agents can be physical, chemical, biological, or psychological.

1. Physical (energy or matter): heat, radiation, electric current, vibration, and sharp, perforating or abrasive objects
2. Chemicals: dust, fumes, vapors, gases, chemicals in general
3. Biological: viruses, bacteria, protozoa, substances produced by animal or vegetable
4. Psychological: bullying, moral harassment, gossip

Let us define some variables related to harmful agents. The ability to cause harm is the harmfulness. In chemical agents, it is the toxicity, and in viruses, the virulence. An agent has more harmfulness than another if under the same conditions, it causes greater damage. Harmfulness depends on the agent and the injected amount.

Solely the dose determines that a thing is a poison (*Sola dosis facit venenum*).[2] Potent poisons may be great remedies. Small doses of strychnine are medicine for the optic nerve.

Absolute dose or dose is the amount injected, ingested, or absorbed by the target. The poisonous dose results from three factors: agent harmfulness, exposure time, and target susceptibility.

The specific dose is the amount absorbed per unit of area, volume, or mass. The specific dose correlates better with damage than the absolute dose. The alcohol's absolute dose (grams or milliliters) is less important than the specific dose (grams or milliliters per kilogram of body weight).

The incidence rate is the amount of harmful agent applied by the aggressive agent in the unit of time or per application. Quantity/time, such as rem/h and kcal/h, measures the absolute incidence rate for aggressive agents of continuous action. The specific incidence rate units are rem/m^2·h and kcal/m^2·h. Quantity/application, such as ml per bite and grams per inhalation, measures the incidence rate for aggressive agents of discontinuous action (venomous animal, toxic gas).

Absorption rate is the amount absorbed by the target (g/s, rem/h, and ml/h).

Contrary to the general feeling, the most toxic products are not artificial, but certain natural products. For a man of 70 kg, the toxic dose of the botulinum toxin is 2 ng, while that of strychnine is 39 mg and that of sodium cyanide is 700 mg.

Susceptibility is the predisposition to suffer damage. Susceptibility depends on the target and harmful agent. We say that the target is susceptible to the agent.

The various parts of a target may have different susceptibilities. The stomach is susceptible to the acetylsalicylic acid (aspirin); the central nervous system, liver, and kidneys are susceptible to chlorinated hydrocarbons.

Aggressive agent's properties and their form influence the susceptibility. Human hearing aid's susceptibility depends on noise form (continuous, intermittent, impact) and frequency (higher susceptibility at 1,000 Hz).

10.10 AGGRESSIVE AGENT AND VULNERABILITY

An aggressive agent is any entity capable of causing damages and losses, injecting harmful agents or preventing their elimination. Wasp injects venom; humidity prevents the elimination of heat from the body. Aggressive agents are harmful agents (toxic gas cloud, X-rays) or vehicles of harmful agent (wasp, snake). The term "aggressive" indicates their quality. When they perform the aggression, we call them aggressor agent.

Aggressive agents can be physical, chemical, biological, psychological, or ergonomic. The physical agent can be mechanical, electrical, thermal, or a wave.

An elastic agent contains elastic potential energy as in compressed spring, taut rod, pressurized gas, or pressurized liquid. A gravitational agent has gravitational potential energy as a body located above or below a target. The potential energy exerts aggressive action when turns into kinetic. When a person falls from a place at height, things happens as the ground had kinetic energy.

The kinetic agent has motion energy as in cars, the moving part of a dynamic equipment, flying objects generated by explosions, and a high-velocity projectile.

Safety Concepts

The kinetic energy results from the transformation of elastic, gravitational, chemical, or electrical energy.

The electric agent has electric potential different from the target. It exerts aggressive action when causes electric current to pass through the target. The electric power may be in the electrostatic or electrodynamic form.

The thermal agent has a temperature different from that of the target. It acts by direct contact, transmitting heat by conduction or convection, or at a distance, by radiation.

Biological agents are vegetal or animal living organisms; they attack by mechanical action (bumps, bites) or by releasing chemicals (acids, caustics, and proteins), biological (enzyme), or other organisms (bacteria, protozoa, viruses).

Chemical agents attack by performing chemical reactions or generating other types of energy. They are explosive, corrosive, flammable, or toxic.

Wave agents: vibration, noise, electromagnetic field, ionizing radiation (including atomic particles), and nonionizing radiation.

Ergonomic agents strike through the relationship with the person. In fact, they are promoters of aggressiveness. Lighting, heavy load, computer screen, or keyboard only promotes the aggressive reflexive action exerted by the person himself.

According to the action over time, the aggressive agents can be (a) agents of continuous action, such as the noise emitted by electric motor, the radiant heat emitted by furnace, vibration of continuous operation machine, and deficient illumination; (b) intermittent, discontinuous, or batch action agents, such as noise of the impact produced by pile drivers; (c) sporadic acting agent, such as the pressure generated by explosion and toxic gas leaking.

The degree of hazard or hazardousness of the aggressive agent results from some properties, such as identifiability, mobility, expansiveness, aggressive energy, aggressiveness, and aggressive capacity. Mobility is the ability to move. Inorganic and vegetal agents are immobile while animals have great mobility, such as the flying insects. Expansiveness is the ability to occupy portions of space. Gases and vapors have high expansiveness, occupying all available space, while liquids occupy the surfaces through which they can flow and the solids have almost null expansiveness.

Aggressive energy is the property that gives the ability to penetrate targets. Gravity field imparts gravitational potential energy; compressed springs, tensioned bodies, and pressurized fluids contain elastic potential energy; bodies in motion contain kinetic energy. Temperature imparts thermal energy, while electric charges impart electric energy and atom bonds impart chemical energy. The biological energy is inherent to living beings and substances produced by them, and the wave energy is inherent in radiation, light, nuclear, and sound.

Without aggressive energy, the agent is harmless. A hammer placed on the ground cannot cause damage. However, falling from a platform acquires kinetic energy to kill a person. After traveling 2,000m, a common rifle's bullet does not have the energy to kill a man.

Potential energy is not aggressive such as the gravitational and elastic energy and the chemical energy of flammable liquids. To attack, the agent needs to transfer energy. The transfer requires a different energy level between agent and target. Car's kinetic energy is not aggressive to the man who is inside, for he has the same speed and therefore the same energy level.

Aggressiveness is the "willingness" to attack, a need to release energy. In animals and humans, it depends on biological and psychological factors. Passive agents have null aggressivity but may be vehicles of harmful agents, such as a plant that contains toxic chemical, causing damage when ingested.

The magnitude of the potential aggression defines the aggressive capacity. Aggressive capacity results from the amount of agent and energy level. Temperature determines the thermal energy level; the electric potential, the electrical energy level; and the velocity, the kinetic energy level. Let us consider some typical cases:

1. The electric lighter for cookers has high voltage, but the small electric charge gives it negligible aggressive capacity.
2. The spark of a lighter has a high temperature, but the small mass of the incandescent particle gives it negligible aggressive capacity.
3. A small dog has low aggressive capacity despite the high aggressiveness.

A rock of one thousand tons at 38°C has a significant amount of thermal energy. However, its aggressive capacity is low (in the thermal aspect).

The vulnerability is the target's weakness toward the aggressive agent. The target may be vulnerable at certain parts and invulnerable in others. Greek mythology provides us with an example. Achilles, a Greek hero, was invulnerable, except at the heel, where the arrow injected the venom that killed him. An arrow is a mechanical aggressive agent. It can inject a chemical or a biological harmful agent.

Elephants are invulnerable to venomous snakes since they cannot perforate their hard thick skin. However, they are susceptible to the venom (harmful agent). Armadillos have the same protection. Pigs are susceptible and vulnerable. However, their defense system has a barrier, the fat layer under skin, which retards venom from entering the bloodstream giving time to the metabolism to destroy it.

10.11 SOURCES OF AGGRESSIVE AGENTS

Source is a system, installation, equipment, material, or activity that contains or is capable of producing or releasing impactful agents. A source may be contained in another one. The one that releases impactful agents is an immediate source and those who feed it are contributing sources. When the target is environmental, we use the denominations receptor body, pollutant source, and pollutant, for the target, the source, and the impactful agent. Let us focus on aggressive impactful agents. Examples are as follows:

1. An industrial furnace in maintenance is an immediate source of contaminated refractory bricks (chemical agent).
2. The fire is an immediate source of soot, thermal radiation, and toxic gases. A leak is a contributing source of fuel (chemical agent), and the atmospheric air is a contributing source of oxygen (chemical agent).

Safety Concepts

3. Radioactive particles are immediate sources of ionizing radiation (physical agent).
4. A solution of sodium sulfide is an immediate source of sodium sulfide (chemical agent). It can also be an immediate source of hydrogen sulfide (chemical agent). The tank that contains the solution is a contributing source.
5. The Sun is an immediate source of ultraviolet radiation (physical agent).

Aggressive power is the number or amount of aggressive agents that the source can release in the unit of time (Kcal/h, mg/day, particles/s, insects/day).

10.12 AGGRESSIVE ACTION FIELD

Action field is the region of space where the aggressive agent acts. It can be a current or a potential action field. A variable linked to the aggressive capacity measures its intensity. Full-field characterization requires the description of the intensity in time and space. The characterization of aggressive fields generated by hazardous events requires consequence analysis techniques. The explosion of 1 ton of trinitrotoluene produces an overpressure of 15 psi (1 bar) at 30 m distance. Therefore, the intensity of the aggressive field at a 30-m-radius circumference centered on the explosion is 15 psi.

Aggressive field is the region where the aggressive agent can cause damage. Outside its limits is the nonaggressive field where the aggressive agent cannot cause damage. The action over time and space classifies the aggressive fields.

1. Action over time

   In a permanent action field, the agent is always present, such as radioactivity and gravity.

   The agent is not always present in an intermittent field. Aggressive agent's presence can be deterministic or random. The frequency may be considerable, such as cars on streets and pilling machine noise, or low, such as the hurricanes.

2. Action in space

   In a continuous action field, such as noise and radiation, the agent is present at all locations.

   In a noncontinuous field, such as cars and animals, the agent is not present at all locations.

   In a uniform action field, the agent is present at all locations with the same intensity, such as atmospheric pressure or gravity at the same height.

   The action field concept may classify the hazardous and unhealthy conditions foreseen in the regulations of many countries. The hazardous area is the area where the aggressive agent has sufficient aggressive capacity to cause acute damage. A borderline from which the aggressive agent has insufficient aggressive capacity limits the hazardous area. Unhealthiness area is the place where the action field intensity can cause chronic damages. The unhealthiness condition depends on both field intensity and exposure time. For each intensity, the exposure time defines the unhealthy condition. Above a certain intensity, the condition may be unhealthy for any exposure time.

10.13 EXPOSURE

Exposure is the presence of a target in the aggressive field. Five exposure categories may occur.

Category I
Target is present where a container has an aggressive agent. Loss of containment would create an aggressive field and the target would be in. Therefore, a hazardous condition or hazardous situation is present, such as in the following situations:

    a. An operator works where a pressure vessel contains toxic gas.
    b. A biologist works where a cage contains venomous snakes.

Category II
Target is present where a container has an aggressive agent, loss of containment occurs, but the aggressive agent is not acting on the target. Therefore, a dangerous situation is present and the target is in danger, such as in the following situations:

    a. An operator works near a pressure vessel containing toxic gas when a leak occurs.
    b. A biologist works near a cage containing venomous snakes when a snake escapes.

Category III
Target is present where a container has an aggressive agent, loss of containment occurs, the aggressive agent acts on the target, but he uses a protection system. Therefore, a hazardous condition or hazardous situation is present, but no damages occur, such as in the following situations:

    a. An operator works where a pressure vessel contains toxic gas when a leak occurs. The toxic gas involves the operator, but he is using a self-contained breathing apparatus.
    b. An operator fitted with a gas mask enters the area containing toxic gas to close a valve stopping a leak.
    c. A biologist works where a cage contains venomous snakes, and a snake escapes and attacks the biologist, but he is wearing safety boots.

Category IV
The aggressive agent acts on unprotected target. The event is damaging, such as the cases mentioned in exposure Category III, but the operators and the biologist are unprotected.

Safety Concepts

Category V

The aggressive agent acts on the target, but its aggressive capacity is insufficient to cause damage, such as a work environment containing toxic gas, but at a concentration lower than the tolerance threshold.[3]

Regarding target's participation, exposure may be either active or passive:

Targets' participation is passive when the aggressive agent escapes and extends its action field reaching targets. To prevent exposure one must impede targets staying in intermittent aggressive fields. When this is not possible, the emergency control system must provide resources for evacuation.

Targets' participation is active when they invade the aggressive field. In some cases, they may invade the containment system itself, such as a person who enters the area isolated for running gamma ray tests.[4] Alarms and isolation can prevent the exposure.

An anti-exposure system is the one that prevents, reduces the likelihood, or lowers the exposure category, such as alarms, rescue systems, isolation, evacuation, and protection.

10.14 ALARM

An alarm is information about dangerous situations. It can be natural or artificial. The color of a metal at high temperature, the rattlesnake's rattle, and the irritant odor of sulfur dioxide are natural alarms. Carbon monoxide, an odorless toxic gas, and some noiseless venomous snakes are treacherous agents. In the absence of natural alarms, or when they are ineffective, the control system must install artificial alarms such as warning signs, siren, and low-odor perception-threshold substances.

Alarms can be permanent and emergency. A warning plate and the color of venomous animals are permanent alarms, while fire siren and the rattling of the rattlesnake are emergency alarms. The emergency alarm system has sensors. An indicator that may be the aggressive agent itself, such as gas concentration, or an element that correlates with it, such as the behavior of some animals alerting other animals about the presence of predators, recognizes a danger.

Alarms are not sufficient to prevent hazardous exposure. Detecting and interpreting alarms are as important as the alarm itself. Only special devices can detect some aggressive agents, such as electricity and ionizing radiation.

Humans and animals' sense organs detect aggressive agents. Equipment replaces them in artificial systems. The probability of exposure increases when, in addition to the absence of natural alarm, the aggressive agent has camouflage, such as the animals whose color resembles that of the environment. Disarray and excessive signaling create camouflages in the workplace, such as poison stored in a bottle of soda. Prior information about the aggressive agents and detectors may prevent exposure, but they do not replace good practices.

10.15 PROMOTERS AND INHIBITORS

Promoters are agents that contribute to increasing agents or targets' properties or any variable's level involved in their relations. In turn, the inhibitors or reductors contribute to reducing the same properties or variable's levels.

1. *Promoters and inhibitors of aggressiveness, aggressive capacity, or harmfulness*
   Water increases the aggressiveness of the concentrated sulfuric acid against aluminum; water reduces the aggressiveness of diluted sulfuric acid against aluminum; the volume promotes water's aggressive capacity against populations, and the bile promotes pancreatic juice's harmfulness against the pancreas.[5]
2. *Promoters and inhibitors of vulnerability or susceptibility*
   Influenza viruses promote people's susceptibility to bacteria, the low temperatures promote steel's susceptibility to mechanical shock, and moisture promotes man's vulnerability to electricity.
3. *Promoters and inhibitors of exposure*
   Winds can promote or inhibit the exposure of people to toxic gas clouds.
   Some people act as virus carriers, promoting exposure of other people.
   The insect known as barber is an aggressive agent, acting as a vector and promoting people's exposure to the protozoan *Trypanosoma cruzi*, a harmful agent that causes the Chagas disease.
   High voltage safety sign inhibits people's exposure to electricity.
   Rattlesnake's rattle and the color of a metal at high temperature are inherent exposure inhibitors.
4. *Promoters and inhibitors of failures*
   Heat and noise promote carelessness failure.
   A device that alerts or prevents an incorrect operation is a failure inhibitor. In the vehicle panel, a light indicates the parking brake applied.

10.16 CONTAINMENT AND RETENTION

A containment system or containment aims at preventing the release of aggressive agents. It can contain the aggressive agent or the entire source and may be mechanical, such as a cage containing a tiger or a pressure vessel containing a flammable product. The structure supporting equipment, materials, people above the ground level, and the rails of a railway line are mechanical containment. In the sodium sulfide solution, the sodium hydroxide or the sodium is the chemical containment of the aggressive hydrogen sulfide.

The same source or aggressive agent can have more than one containment system, and a release does not occur unless all the containment systems fail. Fuel oil may have a steel tank as the first containment and a concrete dam as the second.

In special occasions, it is necessary to open some containment systems and for this, they have doors, while others keep permanent openings to let pass the nonaggressive agents. Doors are weaknesses since someone may open them by mistakes. Containment fails by rupture or opening doors. To increase the reliability, doors have locks. Door and lock have broad meaning here. In cages, they are actually doors and locks, while in pressure vessels, the door is a drain valve, and the lock is the device that impedes opening the valve.

Restoration and/or protection systems prevent containment failures.

The retention system or retention aims at preventing the passage of the aggressive agent. To avoid the unduly retention of other elements, it must be selective. The

Safety Concepts

retention percentage or the aggressive agent amount at the exit expresses the retention performance. Monitoring the retention performance prevents the undue passage of aggressive agent.

The retention system acts by actual retention (filters and cyclones), transformation of the agent into a less aggressive or innocuous one (combustion and biological oxidation), or reduction of the energy of the aggressive agent (noise damper).

Filters, cyclones, and decanters retain mechanically; mask filters against toxic gases, flue gas scrubbers, and car exhaust catalysts retain chemically; electrostatic precipitators retain electrically; separators of ferromagnetic impurities retain electromagnetically; and effluent treatments retain biologically.

The retention system is a combat system since it acts on aggressive agents.

10.17 RUPTURE AGENT

The rupture agent breaks, cancels, destroys, or disables the containment system or opens its doors. A pressure vessel is a mechanical containment of the flammable liquid, and corrosion is a rupture agent. The sodium of sodium hydroxide is the chemical containment of the hydrogen sulfide in the sodium sulfide, and the sulfuric acid is a rupture agent. In fuels, chemical bonds are energy containment, while oxygen and temperature are rupture agents. Containment systems resistant to certain rupture agents are vulnerable to others. Copper is resistant to hydrochloric acid but fragile to ammonia.

10.18 ISOLATION

Isolation systems, such as rope, barrier, wall, and fence, prevent targets entering the aggressive field. Isolation systems, such as valves, spades, spectacle blinds, and flame arresters, prevent hazardous interactions, ingress of aggressive agents, or hazardous promoters. Some isolation systems or devices, such as nonreturn valves, only permit passing in one direction and block the reverse passing. Some systems have a dual function, acting as containment and isolation systems. The sander's protection system contains the disc fragments in case of breakage and prevents the man from touching it. Therefore, it is not a machine guard, but isolation and containment.

Impenetrability or penetration resistance is the fundamental property of the isolation systems. The more the impenetrable, the better the isolation. Some systems exert good containment, but poor isolation. A layer of air constitutes a good containment for electric charges, and a man could work at a short distance from an electric bar. However, the air is a null mechanical isolator and, at the slightest neglect, the man enters the aggressive field.

10.19 RESTORATION

Demand is the event that, in the absence of emergency control actions, evolves to higher level stages of the process that produces damages. The actions of rupture agents are demands. Corrosion is a common demand in industrial units.

Restoration systems recover the normal state of the containments. Restoration may use three strategies:

1. Detecting containment system's changes and mobilizing restoration services (inspection of pressure vessels);
2. Neutralizing the rupture agents (cathodic protection);
3. Putting another containment system into operation (storage tank containment basins).

The rupture agents themselves can activate some restoration systems such as the centrifugal force that triggers a rotating machine shutdown device to protect it against over speed. Others require complex systems with demand sensing element, information, and control action, such as inert gas injection maintaining a reactor's fuel/oxygen ratio outside the flammable range. Demand is the failure of the flow of inert gas. The restoration system has a sensor of inert gas flow, with low-flow alarm and controller action closing the fuel valve.

10.20 COMBAT

Aggressive agents can cause harm or become rupture agents. The combat system acts on aggressive agents or their sources eliminating them or reducing their aggressive capacity. Both emergency and normal situations demand combat. When the aggressive release is inherent to the process, combat is a normal operation performed by air conditioning system (heat), local exhaust ventilation (toxic gas, dust), diluting ventilation (toxic gas, dust), and water mist (dust, vapors). If the aggressive release results from hazardous events, combat is an emergency activity, such as fire extinguishers, deluge systems, and gaseous fire protection and suppression systems.

10.21 PROTECTION

Protection systems interpose between aggressive agents and targets to avoid the category IV exposure. They protect vulnerable points of the target. Protection can be stationary or mobile, natural or artificial, individual or collective, permanent or emergency. Paint is a stationary protection widely used to protect containment systems against chemical aggressive agents (rupture agents). Passive fire protection protects equipment against fire and blasts effects. Crash barriers protect equipment against mechanical shocks. Skin fat and melanin formed by exposure to the sun are natural protectors, while creams and clothing are artificial ones. Glasses and masks are personal protective equipment (PPE); creams and pastes are personal protection materials. Shelters, bunkers, and water curtains are collective protectors. Dangerous and unhealthy operations require protection. Machine enclosure for noise reduction, called collective protection, is actually a containment system.

Protection systems act within certain limits defined in their design, construction, and tests. The aggressive agents, their intensities, or concentrations define the criteria to assure the desired performance.

Safety Concepts 143

When it is necessary to retain aggressive agents but letting pass the beneficial or innocuous ones, the protection system has doors fitted with retention systems such as the gas mask, which contains a chemical or mechanical filter.

10.22 DEFENSE

Emergencies cause damages and losses; defense systems minimize them. The defense can be external or internal to individuals. External defense systems prevent the target exposure (isolation, protection), remove others from the aggressive action field (evacuation), combat aggressive agents (combat), and rescue the victims (rescue).

Internal defense combats the injected harmful agents, either transforming them into innocuous ones or consuming their energies.

Capacity and selectivity are key internal defense's parameters. Capacity is the amount of harmful agent that the defense system can successfully process at unit time. Selectivity is the ability to fight harmful agents without harming the organism.

Defense system changes the variables that characterize the community normal state. A community affected by toxic gases activates the civil defense and the firefighting group. The overall emotional state remains unstable until the defense acquires the control of the emergency. When the defense eliminates the harmful agent, the variables return to the normal state. However, they may remain outside the normal range because the combat neutralized the harmful agent, but did not eliminate it. The defense maintains compensatory actions for the disturbances. Similarly, when a virus strikes the human body, the immunologic system increases body temperature and the number of antibodies. Normal state returns when the defense neutralizes the aggressive virus.

The defense system may produce harmful agents. Let us consider the following cases:

1. Heat-induced diseases such as dehydration, cramps, dizziness, and fainting result from the physiological response attempting to keep body temperature constant;
2. When the immune system fights bacteria, one of the altered variables is the number of antibodies, which may attack the body itself, causing rheumatism or heart disease;
3. When man ingests methanol, the formic acid produced by the oxidation for elimination is much more harmful than the Methanol itself.

An interesting case occurs in companies. Managers suffer more organizational "pressure" (aggressive agent) than their subordinates do. However, they have better conditions to eliminate stress (harmful agent).

10.23 RECOVERY SYSTEMS

The recovery systems, such as medical assistance, environmental recovery, and assets reconstruction, aim at stopping the evolution of events that lead to damages, recovering the targets hit by aggressive agents, and returning, if feasible, their original state. Damages and losses depend on aggressive agent's aggressive capacity, harmful agent harmfulness, protection, vulnerability, susceptibility, target defense system, and exposure time. When the aggressive action terminates, the damage can achieve a steady state or continue to evolve. The recovery systems of living beings act as promoters since the organism itself does the recovery.

10.24 HAZARD AND QUALITY

Let us consider a vendor V and a consumer C. C wants to buy an input that performs functions to satisfy his needs. V produces a product P that performs those functions. Therefore, the same object is a product for V and an input for C.

Useful functions interact with the needs; aggressive functions interact with vulnerabilities. It may occur that some functions satisfy needs, other ones satisfy needs partially, some have no needs to satisfy (useless), and some needs have no functions that could satisfy them (unsatisfied). Aggressive functions may affect vulnerabilities, and the consumer may perceive functions that actually do not exist.

Quality arises from product/consumer relations. Congruence between needs and functions generates C's satisfaction, which measures P's quality level. P's ability to perform useful functions for C promotes its quality.

Functions include reliability and availability; needs include time and place. A refrigerator may have a high-quality level at Ecuador, but it is useless at the North Pole although its functions remain the same. Similarly, a high-quality wool coat is useless at Ecuador.

P's ability to perform aggressive functions, providing harmful inputs, generates hazards. Hazard arises from the relationship between aggressive agent and the target. Hazard is the negative face of quality and is a price of the product. Figure 10.1 shows the results of the relationships between a consumer and a product.

The hazard concept is, along with that of risk, the most important basic concept. Let us make an analogy between hazard and heat. Absorbing heat, the thermal energy of the object increases. By analogy, let us say that an agent, substance, situation, or event contains harmful energy. A high amount of harmful energy implies high hazardousness level. Just like thermal energy, it is impracticable to remove the harmful energy at all. The harmful energy results from aggressiveness, aggressive capacity, mobility, expansiveness, harmfulness, and the absence of natural alarms. However, like heat, the harmful energy needs to transit to cause harm. Therefore, analogous to heat that is thermal energy in transit, hazard is harmful energy in transit.

Hazards identification includes hazardous agents, hazardous substances, hazardous mixtures, hazardous interactions, hazardous animals, hazardous or dangerous situations, hazardous operations, and hazardous events. As the hazard does not exist absent from the relation between aggressive agent and target, it usually refers to people. High hazardousness does not imply necessarily more damage over time since damages depend on exposure and control systems, i.e., it depends on the risk, the object of Section 10.29.

Safety Concepts

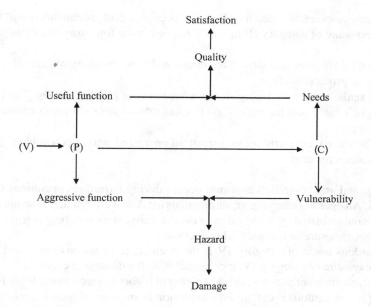

FIGURE 10.1 Relationships between a consumer and a product.

10.25 HAZARDOUS EVENT

An undesirable event is the one that causes losses of any kind, including those arising from low productivity or quality as delays, absenteeism, strikes, and the shutdown of industrial units. A hazardous event is the undesirable event that releases or generates aggressive agents and exposure targets to aggressive fields. Setting fire to a gas burner or flammable liquid for firefighting training or entering areas containing unrestrained aggressive agents, wearing the suitable PPE, do not constitute hazardous events. Those activities are operations and hazardous works.

Five hazardous events of different categories may occur.

Hazardous event of category 0 is an event in the sequence leading to loss of containment. Cooling water loss and shutdown valve failure are hazardous events of category 0.

Hazardous event of category I is the occurrence of unwanted and unscheduled exposure of category I. The target penetrates the aggressive field when the aggressive agent is in the containment system. A visitor inadvertently enters the area where a leash restrains a ferocious dog; an operator enters an area isolated for radioactive tests when the source is inside an armored box, but the test is about to start.

Hazardous event of category II is the loss of containment, or the unscheduled, unwanted, and unnecessary entrance of a target in a discontinuous action field containing an unrestrained aggressive agent. If the action field is continuous, a hazardous event of category III or category IV occurs. Flammable and toxic material leaks and fires are the hazardous events of category II of the greatest interest in chemical industries.

Hazardous event of category III is the occurrence of unscheduled and uncontrolled exposure of category III as would happen in the following situations:

1. Hot oil splashes and strikes the operator during sampling of fuel oil, but a mask protects his face.
2. A snake strikes the biologist, but a leather boot protects his leg. The biologist's entry into the snakes' pit is a hazardous operation with exposure of category II.
3. The electrician's arm leans against an energized wire, but a rubber glove protects his arm.

Programmed and controlled exposure occurs during hazardous operations of category III. An operator carrying an autonomous mask and air cylinder assembly enters an area contaminated by toxic gases to close a valve, stopping the gas leaking. The hazardous exposure of category III is necessary.

Hazardous event of category IV is the occurrence of an undesired and uncontrolled exposure of category IV. It coincides with the damaging event.

A hazardous event may evolve into another of higher hazardousness level. Fuel gas leaking is a hazardous event, and the explosion is an event of higher hazardousness level. Hazardous operations may evolve into hazardous events when failures occur.

10.26 HARMFUL AND DAMAGING EVENTS

A damaging event is the occurrence of category IV exposure. It is the final event to prevent. An agent acts on unprotected target when all controls fail. Damages are immediate consequences whose dimensions depend on agent's aggressive capacity, target's vulnerability and susceptibility, exposure time, and random factors. Any event without damage, but causing moral harm such as car theft, is a harmful event. Most hazardous events are damaging or harmful. The classification in one or another category depends on the target. A fire is a hazardous event for people. If it is their property, is also a harmful event, while the flame incidence is a damaging event for them. For the property that burns, the fire is a damaging event.

10.27 EMERGENCY

State of emergency or emergency is the manifestation of hazards. Risk factors emerge from the virtual to the real world generating damages and losses. Damages and losses result from a successful undesirable chain of events. When the first event occurs, an initiator event or a demand, the emergency starts.

Emergency control includes detection, mobilization, and intervention. Intervention includes restoration of containment, combat, and defense. An emergency control organization is the most complete emergency control system containing operational and organization systems, equipment, installations, and men.

Generally, emergencies begin before the initiator event. In fact, it begins when the normal state of any component of any system starts changing. Anatomical and physiological changes are still undetected, but an abnormality exists. A chest pain

Safety Concepts

reveals a heart attack that started when fatty deposits began reducing the diameter of coronary arteries.

Anatomical changes remain hidden due to lack of techniques or instruments or by the impossibility of performing inspections. Physiological changes remain hidden when the workload is lower than design load or the alterations are undetectable by the available means.

10.28 ACCIDENT—INCIDENT—NEAR MISS

An accident is the abnormal occurrence that contains a damaging event. Damages and losses, however negligible, always occur.

An incident is an abnormal occurrence that contains a hazardous or unwanted event, but no damage. Random factors or control systems prevent the hazard's successful progress. Hazardous events of category 0, III, and I are incidents. Hazardous events of category II are also incidents when only productivity losses occur. The damages, not the losses, differentiate accidents and incidents. The following are examples of incidents:

1. A person enters the isolated area during preparation for gamma ray test. Safety personnel detects the fact in time, and the event affects only productivity.
2. A lion escapes from zoo's cage. Safety personnel recaptures him, and the event only scares the visitors.
3. An operator is taking hot oil samples when the liquid splashes, hitting the protective equipment, but no injury occurs.
4. A control valve fails, cutting the injection of inert gas into a reactor, but the fuel cutoff system acts avoiding an explosion, and the event only entails production losses.

A single event may be accident or incident, depending on the point of view. A hammer drops from a platform but does not hit anybody. The event is an incident for the workers but an accident for the hammer.

Near miss is the real or virtual event that *almost* turns into an accident. A worker enters inadvertently the area isolated for radiography, but no exposure occurs. This event is an incident since it is a real fact. However, when someone realizes that was close to performing a hazardous intervention, causing an accident, we have a virtual event, an *almost accident*. No fact, not even an incident to report. The critical incident technique identifies this kind of near miss through interviews.

10.29 RISK

The risk is the expected damage or loss over time. It is a random variable associated with events, systems, installations, processes, and activities. The expected value and the variance characterize the risk probability distribution.

When we approach an object of study, we focus the total risk or the risk associated with a particular aspect. The overall risk is the sum of all risks associated with credible hazardous events.

The real risk is an objective variable although the variability of the estimations is considerable. Perceived risk is a subjective variable because its evaluation results from beliefs, emotional factors, and traumatic experiences. People use subjective methods to assess risks. For this reason, the real and the perceived risks usually show considerable differences. The perceived risk associated with an airplane travel is higher than that of a car travel, while the real risks have an inverse relation.

Risk assessment studies refer to the original risk associated with an object as gross risk. The remaining risk after introducing control measures is the net risk. Individual risk is the individual damage or loss expected over time. Social risk is the community damage or loss expected over time. Usually, negative powers of ten deaths/person-year are risk units. Let us consider 10^{-7} deaths per person-year as the individual risk for people residing next to a chemical industry. If it were possible to eliminate all other causes of death, people who live next to the industry would live 10^7 (ten million) years.

The risk associated with a hazardous event results from event's frequency and consequence. Frequency and consequence are risk factors, and the following expression defines risk mathematically:

$$\text{Risk} = \text{Frequency} \times \text{Consequence} \qquad (10.1)$$

Events/year and events/h are usual frequency units. Damage/event and loss/event are usual consequence units. Damage/year, loss/year, dollars/year, deaths/year, and deaths/person-year are usual risk units.

The risk associated with a hazardous event depends on the event and the scenario, which includes location, targets, exposure promoters, and emergency control systems. Therefore, a hazardous event of serious consequence may have a low associated risk if the frequency of occurrence is low. On the other hand, an event of lesser consequence may have a high associated risk if the frequency is high. Therefore, risk comparisons require knowledge of the two risk factors.

Frequency and consequence result from risk factors. Rupture agents, aggressiveness, mobility, expansiveness, and alarms are frequency factors. Aggressive capacity, harmfulness, vulnerability, and susceptibility are consequence factors.

Let us consider two events: a coral snake bite and a rattlesnake bite. Comparing consequences, the coral snake bite is more severe; however, the frequency of rattlesnake bite is much higher, resulting in higher risk.

The risk and emergency control measures can reduce the action of all factors taking the risk to an *As Low As Reasonably Practicable* level.

10.30 MAINTENANCE

Maintenance function aims at recovering the normal state of a component so that it keeps performing its mission with suitable performance and reliability. Maintenance consists of detection and intervention. Detection consists of inspection and testing.

Non-stimulated inspection is a simple observation. Stimulated inspection applies a stimulus and observes the outputs. Inferences on component state come from the outputs. Applying internal pressure and observing the mechanical behavior of a

Safety Concepts

pipe, increasing flow and observing pump vibrations, applying penetrating liquid and observing the plate cracks, subjecting a welding to X-rays and observing the radiography, applying physical exercises, and observing the heart electrical signals are stimulated inspections.

Regarding strategies, maintenance can be corrective, preventive, or predictive.

Corrective maintenance consists of intervening when failures reveal abnormalities. Usually, it is not a strategy but a necessary intervention when a fault state occurs. If the failure has hazardous consequences, this strategy may cause damages and losses. Therefore, it applies to noncritical components, or the ones that present failures not expected by another strategy. Lamps are a typical example.

Preventive maintenance consists of intervention based on operation time or demand frequency, replacing or inspecting components independent of their behavior. This strategy applies to critical components of well-known reliability, and no inspection or test can assess its current state. Replacement of components that can fail due to fatigue, such as screws and cables, is a typical case. The investigation of aircraft accidents reports an emergency with an airplane whose structure ruptured due to fatigue. The cause was the high number of landings and takeoffs, while the controlled parameter was the total miles. Another factor to consider is the possibility of introducing unwanted changes or a component having a failure rate higher than the replaced. The risk is higher when the component is complex.

Predictive maintenance intervenes to repair or replace non-faulty components when indirect signs indicate alterations in its state. This strategy has two advantages: the first is to avoid early interventions, and the second is to avoid postponing an intervention beyond the instant from which failures are likely to occur. The predictive maintenance is suitable when noninvasive observations can predict the component state, such as the vibration measurements that indicate the state of centrifugal pump bearings.

Predictive and corrective strategies define interventions through different ways. The predictive maintenance intervenes when alterations still not threaten the normal operation, and detection requires special methods or tools. In turn, the corrective maintenance intervenes when alterations start affecting the normal operation, and the abnormality signs are visible.

10.31 OPERATION

Operation function interacts with systems, starting, stimulating, monitoring, altering functional spectrum, and stopping. Its deployment leads to the auxiliary functions, start-up, operate, and shutdown. Petrochemical plants and nuclear power plants operate continuously for long periods, requiring virtually no human intervention while batch processes have a high start-up/shutdown frequency, and man is an essential process component (man–chainsaw, man–crane, and a pilot during landing and takeoff).

Batch processes have higher associated risks than continuous processes of equivalent hazard level. The increased risk stems from the high frequency of man/machine interaction. Moreover, it is precisely in those phases that significant variations in temperature and pressure occur, increasing leak frequency and the likelihood of formation of explosive mixtures. On the other hand, the start/stop low frequency in continuous operation processes impairs accumulation of experience.

Operation under normal conditions consists of monitoring and intervention. Operational monitoring focuses on operating conditions, functions, performance, environmental conditions, requirements of other systems, and external aggressive agents. Deviations from design conditions can be aggressive, promoting failures and acting as rupture agents.

System reliability depends on operating and environmental conditions. The operating conditions include load, characteristics of the inputs, frequency of stops and starts, temperature, pressure, and speeds. Environmental conditions include humidity, temperature, atmospheric pressure, presence of dust, and other aggressive agents.

Operation in emergency conditions requires monitoring, intervention, and shutdown under tension and a short time to analyze and decide. Even in automated systems, the man performs functions non-transferable to equipment. When equipment fails, human action is the last resort.

10.32 PROJECT

The project function implements new systems, designing them to meet needs. Good project analysis uses value engineering, hazard identification, and risk assessment. The method promotes functions identification and develops alternatives to improve project value. Value engineering imparts safety a holistic approach, and safety promotes the identification of aggressive functions.

Design phase selects processes and establishes maximum allowable noise levels, material aggressiveness, operating limits, utilities, and maintenance facilities.

The design phase is the most economical phase to control risks. When the solution of a problem requires a monetary unit in the conceptual or basic design phase, detail design will require 10; construction and assembly phase, 100; and operational phase, 1,000. If a serious accident occurs, it will require 10,000.

In addition to the project itself, the implementation of new systems involves acquisition, conditioning, construction, and assembly. These functions also require care and risk analysis.

10.33 SAFETY

Safety is a variable inversely proportional to risk. The higher the risk, the lower the safety, and vice versa. Increasing safety means reducing risk. The word safety may denote the safety function or the organization aimed at promoting the risk and emergency control.

Let us examine the relationships between mission, reliability, and safety. Race cars must have high performance, even limiting reliability and safety. Reliable and safe cars that lose races do not survive. An enemy aircraft will shoot down a reliable and safe military aircraft that has poor fighting performance. Commercial airplane reliability is a critical variable but cannot impair the airline company survivability. Therefore, safety is dependent on other vital variables.

Safety function reduces damage and losses caused by aggressive agents. Productivity losses and low-product quality are not the main concerns. However, safety methods and techniques can help to solve productivity and quality problems.

Safety function strategy is to neutralize aggressive agents, but total neutralization is impracticable. The cost system incorporates a tolerable residual risk or distributes it over time through insurance. Moreover, the lower the tolerable risk, the greater the cost to achieve it. The cost of the control measures may derail projects and activities.

Risks may be low because the hazards are of small magnitude or the controls neutralize them. In this case, the neglect of safety function raises the risks.

The safety function has two complementary strategies: preventive (risk control) and corrective (emergency control). The first strategy controls the latent factors, and the second, the hazard manifestations.

NOTES

1. TELES, Maria Luiza Silveira. Aprender psicologia (Learning Psychology), São Paulo: Brasiliense, 1990, p. 39.
2. "Only the dose makes the poison," (Paracelsus).
3. Tolerance threshold is the concentration of a chemical agent or the intensity of a physical agent present in the work environment under which workers can work throughout their working life without adverse effects on their health.
4. Radiography using gamma rays.
5. Gallbladder's stones obstruct the bile duct causing bile reverse flow into the pancreas. Pancreatic juice does not affect the pancreas but activated by the bile, it begins to digest it, causing severe acute pancreatitis.

BIBLIOGRAPHY

1. Capra, Fritjof. 1982. *The Turning Point*. USA: Bantam.
2. Deming, William Edwards. 1990. *Quality: The Management Revolution*. Brazil: Saraiva. (Edition in Portuguese)
3. Lewis, Elmer Eugene 1987. *Introduction to Reliability Engineer*. USA: John Willey & Sons.
4. Lees, Frank P. 1996. *Loss Prevention in the Process Industries*. UK: Reed Education and Professional Publishing.

Index

A

Ability, 2, 14, 21, 23, 26–27, 70, 86, 90, 107, 110, 114, 124, 133, 135, 143–144
Absenteeism, 35, 145
Accident
    analysis of, 4, 87, 89
    and behavior, 118–119
    catastrophic, 54
    causes of, 33, 56, 85, 121, 124
    consequences, 1, 4, 6–7
    fatal, 59, 112
    and human factors, 1
    and incident, 147
    major/serious, 21, 58, 64, 67, 75, 88, 130, 150
    minor, 82, 88
    a multifaceted phenomenon, 3–5, 33, 36
    nature of, 3, 4, 36, 65
    frequency, 1, 88
    outside the organization, 35
    rate, 37, 65
    prevention of, 1, 12, 33, 54, 110, 113
    probability, 121
Accountability, 118
Adherence, *see* organization
Administration, 18–20
    and Chess, 18–19
    and man-apes relationship, 20
    and man-pig relationship, 20
Agent
    aggressive, 4, 7, 11, 19, 37, 39, 41, 44, 46, 53, 55–56, 60, 62–63, 69, 71–72, 86, 89–90, 105–106, 127, 130–145, 150–151
    aggressor, 134
    biological, 135
    chemical, 133, 135–137
    command, 123
    elastic, 134
    electric, 135
    ergonomic, 135
    external, 35, 56, 89–90
    failure-promoter, 36, 90, 139
    gravitational, 134
    harmful, 127, 132–134, 136, 140, 143–144, 146
    hazardous, 56, 144
    impactful, 5, 51, 127, 132–133, 136
    injectable, 132–133
    inoculated, 60
    inorganic, 135
    intruder, 89, 121, 124
    kinetic, 134
    natural, 59
    passive, 136
    physical, 134, 137
    primary, 121–122
    of rupture, 46–48, 58, 69, 73, 86, 90, 127, 141–142, 148, 150
    secondary, 91, 122–123
    thermal, 135
    treacherous, 139
    vegetal, 135
    wave, 135
Aggregate, 51, 53, 87, 99, 128
Alarm, 7, 62–63, 69, 73–75, 94, 107, 123–124, 127, 139, 142, 144, 148
Andersen, K. E. A., 18
Approach, 1–4, 24, 41, 150
Attitude, 1, 8, 20, 83, 85, 112–120
Audit, 8, 36–38
Authority, 9, 12, 21, 52, 113–114, 118

B

Barrier, 44, 48, 53, 63, 69, 127, 136, 141–142
Behavior, 1, 3, 7–9, 12–14, 20, 22, 30, 34, 37, 40–41, 57, 66, 70, 85–86, 89, 108, 112–121, 131–132, 139, 148–149
Belief, 1, 4, 6, 8–9, 17, 20–21, 30, 36, 38, 70, 89, 112–114, 116, 119–120, 128, 148
Blast, 69
Bleve, 26
"Boiled frog", 119
Boilover, 26, 31n7, 107, 125n1
Brainstorming, 75, 77, 103
Bytheway, C., 100

C

Camouflage, 139
Carelessness, 89–91, 108–111, 121–124, 140
Cartesian–Newtonian, 1–2
Cause tree analysis, 88–89
Characters, 21, 113
Charisma, 18, 21
Checklist, 53, 67, 71–72, 75–76, 78–80, 83–84, 86–87, 124
Civil defense, 12, 40, 143
Combat, 7, 39, 41, 43–44, 47, 63, 69, 107, 128, 141–143, 146
Commitment, 9, 27, 70

153

Committees, 8, 24, 34, 36
Common cause failures, 57, 79, 86, 124
Communication, 23, 43–44, 47, 86–87, 89
Community, 11–12, 15, 17, 20–21, 23, 28, 40, 66, 75, 96, 102, 120, 143, 148
Competence, 14, 18, 112
Consequences analysis (CA),75–76, 78, 80
Consumer, 11, 15, 23, 42, 95–96, 99, 102, 131, 144
Containment, 4, 7, 39, 41, 44, 46, 58, 63, 69, 86, 90, 127, 138–142, 145–146
Cooperation, 15, 30
Cousteau, J., 21
Creativity, 5, 14, 26–28, 70, 73, 75, 88, 93
Critical incident technique (CIT), 7, 82–83, 87, 147
Culture, 1, 6, 8–9, 12, 14, 18, 20–21, 23, 25, 27–31, 35, 37–38, 43, 56, 67, 70, 86–89, 110, 116, 118, 120, 123–124, 130

D

Damage, 4–5, 7, 11, 36–37, 39, 41–43, 60–61, 66, 88–89, 95, 98, 107, 111, 127–128, 130–137, 139, 141, 143–149, 151
 expected/potential, 60, 88, 116
 generation mechanism, 43, 69, 89
 moral, 131
 to people, 4, 131–132
 permanent, 61
 to property, 61, 4, 11, 41–42, 61, 131
Danger
 and alarm, 139
 and careless failures, 108–110
 and evacuation, 48
 and exposure, 138
 and risk, 60
Defense, 1, 12, 18, 39–40, 69, 106, 127, 133, 136, 143–144, 146
Deflagration, 60
Demand, 37, 39, 42, 51, 56, 59, 79, 89–90, 141–142, 146, 149
Detection, 7–8, 24, 37, 39–40, 42, 46, 67, 69, 73–77, 83, 85–88, 93, 98, 105–106, 119, 123–124, 127, 130, 139, 142, 146–149
Detonation, 60
Development, 6, 8, 11, 12–14, 18–19, 21, 22–24, 29–30, 36, 39, 42–43, 87, 99, 102–103, 113
Deviation, 7–8, 37, 64, 67, 72–78, 83, 85–87, 93, 107, 111–112, 118–119, 129–131, 150
Disturbance, 64–66, 119, 129, 132–133, 143
Dose, 133–134, 151n2

E

Ego-status, 14
Emergency, 8, 39, 41–42, 142–143, 150
 alarm, 139
 auxiliary functions, 40
 causes and effects, 39
 control, 21, 36, 39–43, 51, 53, 56, 59–60, 71–73, 75–81, 85–86, 90, 105, 113, 139, 141–143, 146, 148–150
 control organization, 6–7, 12, 17, 26, 39, 42–43, 146
 failure analysis in, 105–106
 as a failure promoter, 109–110, 124, 146
 initiator of, 39–40, 51, 127, 146
 location, 40, 47
 management in, 24, 39–40, 44
 medical care in, 44, 46
 resources in, 26–27
 and risk control, 33
 scenarios, 40
 top event of, 39
Environment, 2, 11, 53, 139
 adaptation to, 12, 29
 and aggressive agents, 51, 128, 132, 136
 and attitude, 112
 changes of, 2, 29
 damage to, 36, 41, 42, 128, 131–132
 as a disposal medium, 6, 35–36, 98, 131
 elements of, 5
 and emergency, 40, 42
 impacts on, 7, 52, 76, 103, 133
 interventions in, 51
 management system, 24
 and organizations, 5, 11, 23, 28
 preservation, 11, 24, 44, 77, 99, 102, 129
 protection agency, 54
 recovery of, 143
 resources from, 86
 risk to, 86, 103
 and systems, 66, 128
 and value analysis, 103
 vulnerability, 60–61
 and work conditions, 66, 85–86, 108, 121–124, 139, 150
Escalation, 76, 106
Esteem, 96–97
Event tree analysis, 55, 72, 75–76, 78, 80–82, 85, 88, 107
Experience, 5, 9, 13–14, 18–19, 26–27, 36, 43, 54, 58, 70, 73, 75, 87–88, 107–108, 150
Explosion, 31, 39, 60–61, 75, 82, 85, 134–135, 137, 146–147
Exposure, 37, 41, 55, 57, 60–62, 69, 86, 96, 127, 132–134, 137–140, 142–146, 148

Index

F

Failure, 3–4, 36, 51, 55–59, 69, 71, 74–77, 79–82, 85–86, 88–91, 93, 103, 105–112, 115–119, 121–125, 130–131, 140, 142, 145–146, 149–150
Failure modes and effects analysis (FMEA), 55, 67, 76–77, 106
Faith, 2
Fall, 9n4, 55, 61, 70–72, 88, 90, 134–135
Fate, 3, 21
Fault tree analysis (FTA), 55, 57–58, 67, 72, 75, 78–82, 85, 88–89
Feedback, 27, 65–66, 119, 128–129
Fire, 39, 60–61, 73, 76, 82, 84–85, 108, 131–133, 136, 145–146
    brigade, 6, 17
    chemistry of, 26
    fighter, 11, 43, 113, 143
    fighting, 26, 43, 47, 73, 77, 103, 107, 139, 142–143, 145
    water, 26
Flexibility, 12
Fondness, 6, 8–9, 21–22, 30, 38, 70, 112–114
Fraction dead time, 56, 80
Function, 1–2, 5–8, 11–13, 17, 19–21, 23–28, 30, 33–37, 39–49, 52–56, 66, 70–71, 74, 76, 78–79, 89, 91, 93–105, 108, 110–111, 113, 119, 122, 127–131, 141, 144–145, 148–150

H

Happiness, and survival, 3, 23
Harm, 5, 51, 55–56, 60–61, 67–69, 85, 89, 110, 127–129, 132–136, 140, 142–144, 146, 148
Hazard, 8, 35, 37–39, 53, 55–58, 63, 67, 71–73, 75, 77–79, 81–82, 84–85, 87, 89, 91, 93, 98, 107, 135, 144–147, 149–151
Hazard and operability studies (Hazop), 55, 57, 67, 72–76, 78, 80, 82, 91n1, 93
Heroes, 8, 20–21, 113
Herzberg, F., 14
Hobbes, T., 25
Holistic, 1–4, 11, 14, 23–24, 26, 30, 34, 36, 57, 66, 87, 96, 101–102, 119, 127, 150
Hurricane, 39, 137

I

Impact, 4–5, 7, 21, 44, 51–54, 97, 103, 113, 127, 132–135
Incident, 7, 37, 82–83, 87–88, 147
Indicator, 7, 37–38, 73, 75, 106, 116, 139

Information, 5, 7, 9, 17, 20–21, 26, 41, 44–49, 51, 53, 85, 87–88, 90, 94, 98, 102, 107, 112–114, 119–120, 123, 129, 132–133, 139, 142
Inhibitor, 86, 105, 139–142
Inspection, 7, 36–37, 52–53, 64, 71, 83–88, 124, 127, 142, 147–149
Integration, 2–3, 11–12, 14, 17, 20, 23, 26, 34, 88, 93, 127
Interaction matrix, 67, 84–85
Interface, 53, 109
Intervention, 5, 7–9, 12, 21, 24, 35–36, 39–40, 51–55, 65, 68, 84, 105, 119, 149
Isolation, 7, 44, 47–48, 62–63, 69, 127, 139, 141, 143

K

Klink, A., 113, 125n2
Knowledge, 1, 5–6, 9, 19–21, 26–27, 35, 41, 52–53, 70, 75, 87–88, 98, 107, 109, 112, 114, 116, 121, 124, 148

L

Label, 53, 124
Language, 20
    nonverbal, 7, 111, 120, 123
Law
    "of the least effort", 3
    compliance with, 105
    conscious failure and, 105, 115
    deviation from, 83
    and tyranny, 25
Leader, 6–9, 12, 17–21, 31, 34, 73, 87, 112–114, 120
Leadership, *see* organization
Life, 1, 6, 8, 12, 15, 112–113, 131, 151
    cycle, 8, 24, 34–36, 83, 98, 103, 105, 130
Locke, J., 25
Logistics, 44, 46
Loss, 1–2, 4–7, 11, 19, 27–28, 31, 33, 37, 39, 69, 88–89, 96, 111, 116, 128, 131–134, 143–149, 151, *see also* damage generation mechanism
    of containment, 46, 138, 145
    of control, 64
    and price, 95
    repairable, 131
Luck, 3

M

Maintenance, 4, 6–7, 12, 27, 35, 42, 51–52, 54, 56, 80, 84–85, 122, 124–125, 127, 136, 148–149

Managers, 19, 114, 119, 123–124, 143
Maslow, A., 13–14
Maslow's pyramid, 13, 102
Methodologies, 1, 19, 22–25, 30, 35, 40–41,
    55–56, 58, 71, 73, 76–77, 79, 81, 83,
    84, 86, 88–89, 93, 98–99, 103, 107,
    116, 130, 148–151
Middle Age, 2
Mission, 11, 13, 23–24, 35, 42–43, 66, 90, 96,
    101–103, 105, 108, 148, 150
Mobility, 19, 37, 135, 144, 148
Modern Age, 1
Monitoring, 8, 27, 35–37, 40, 73, 109, 141,
    149–150
Morale, 14, 115, 128, 131
Morphy, P., 18
Motivation, 3, 13–14, 17, 41, 88, 94, 109, 112
Myth, 4, 8, 20–21, 89

N

Near miss, 37, 82–83, 87, 147
Number one, 8–9, 18, 34

O

Occurrence, abnormal, 7–8, 33, 37, 67, 87–88,
    105–106
Omission, 105, 120, 123
Operation, 6, 12, 23, 27, 36, 39, 54–56, 58, 65, 67,
    80, 106, 108, 123–124, 135, 140, 142,
    144–146, 149–150
Organizations, *see also* emergency control
    activities of, 6, 8, 12, 34, 36, 40, 42
    and adherence, 14–17, 23, 129
    and administration, 19
    characterization of, 29, 87, 128, 150
    clients of, 11, 14–15, 23, 95–96, 102
    and climate, 7, 14, 17, 19, 24–25, 28, 87–88,
        129, 143
    and culture, 6, 8, 12, 14, 20, 23, 27–31, 35, 37,
        87–88, 110, 116, 120, 123–124
    development of, 11, 21, 22, 29–30
    and environment, 5, 11–12, 30
    foundation of, 29
    functions of, 17
    holistic, 5, 30, 96
    interventions in, 8, 21
    and leadership, 6, 8–9, 12, 18, 21, 23, 27,
        29–31, 34, 37, 87–88
    and life cycles, 6, 35
    as living systems, 12–13
    and management system, 6–9, 17–27, 29, 31,
        33, 37, 88, 110
    mission of, 11, 13, 96
    organizational field of, 8, 27–29, 89
    organizational systems of, 22–23, 34
    and people behavior, 13–14
    and people needs, 13–14, 25
    performance of, 19, 29
    poles of, 17, 25, 28
    products of, 35
    programs of, 8, 17, 30, 36
    regulatory function of, 6, 12, 23, 27, 89, 91
    resources of, 25–27, 40, 42, 89
    and Risk management, 33–38
    and rites, 17
    safety at, 11
    safety diagnostic of, 7–8, 38, 128
    structure of, 28, 30
    survival of, 23, 101
    as systems, 11, 20, 129 (*see also* systems)
    values of, 6–9, 23, 30, 112
    vision of future, 30
    vital functions of, 93, 102

P

Paradigm, 1–2, 20, 22, 112, 119–120, 130
Perception, 1, 20, 117, 130, 139
Perpetuation, 3, 5, 20
Philosophy, 2, 22, 31
Piper Alpha oil platform, 54
Plan, 7–8, 14, 16, 19, 22–24, 33, 36–40, 42–43,
    53, 64, 67–68, 71–72, 83, 85–88,
    108, 113
Poison, 109, 134, 139, 151n2
Policy, 7, 17, 22–24, 33–34, 38, 40, 43, 70, 89
Posture, 7, 118, 120
Preliminary risk analysis (PRA), 55, 71–72,
    78–80, 82
Prevention, *see* accident
Price, 68, 95–96, 116–118, 144
Principles, 7, 18–19, 22, 30, 33, 40
Probability, 5, 37, 40, 56–60, 66, 69, 79–80, 85,
    88–89, 108–110, 115, 118, 121, 130,
    139, 147
Procedure, 3, 5–8, 22–23, 26–27, 38, 40, 42–43,
    46, 52–53, 62, 68, 70–71, 73–75, 77,
    90–91, 102–104, 107, 109–111, 113,
    115–116, 118, 121–124, 129
Process, 4–8, 12, 18, 22–23, 26, 29, 34–36, 42,
    51, 52–53, 55, 64, 66–68, 70–75,
    77–79, 83–76, 88–89, 99–100, 106,
    108, 114, 118, 120, 128–129, 141–143,
    147, 149–150
Productivity, 6, 11, 19, 23–24, 27, 73, 102, 103,
    110, 129, 145, 147, 151
Program, 6–9, 17, 22, 24, 29–30, 33–34, 36,
    38–40, 70, 73, 89, 119
Properties, 3
    preservation of, 11, 34, 42
    as target of aggressive agents, 40–41, 103,
        106, 132, 146

Index

vulnerability and damages, 60–61, 96
Protection, 7, 41, 47, 54, 58, 63, 69, 90, 115, 124, 127, 136, 138–144
Public, 6, 27, 40–41, 43–45

Q

Quality, 3, 7, 11, 16–17, 19, 23–24, 27, 37, 51, 56, 61, 63, 64, 68, 70, 77, 93–96, 98, 102–103, 110, 117, 129, 134, 144–145, 151
Quantitative risk analysis, 37, 57–58, 60, 67, 79, 81, 86

R

Radiation, 60, 133, 135–137, 139
Radioactivity, 44, 98, 137, 145
Rational, 1, 2, 26–27
Recognition, 14, 23
Recovery, 7, 61, 63, 119, 128, 143–144
Reliability, 40, 51, 64, 68–70, 94, 124, 128, 130, 140, 144, 148–150
Rescue, 7, 39, 41, 44, 47, 49, 63, 127, 139, 143
Resistance
  to changes, 30
  of isolation, 141
  of material, 27, 44
Restoration, 39, 63, 69, 82, 90, 127, 130–131, 140–142, 146
Retention, 141, 143
Retirement, 6, 14, 35–36
Risk
  "As Low As Reasonably Practicable", 148
  acceptable, 66
  of activities, 6, 33–35, 52, 54, 66, 79
  analysis, 7–8, 19, 22, 34–36, 40, 42, 53, 55–56, 65, 67, 71, 76–77, 79, 81, 87–88, 102, 150
  assessment, 35, 55, 57, 61–64, 67, 116, 148–150
  in batch process, 149
  and behavior, 116–119
  biological, 55
  category of, 61–63
  chemical, 55
  concept of, 147
  control, 3, 6–8, 21, 33–36, 52–57, 60–64, 66–68, 71–73, 75–78, 85–88, 150–151
  controller, 53, 65–68
  definition of, 33, 88, 147–148
  diagnosis, 38
  ergonomic, 55
  of facilities, factors, 6–7, 33, 37, 39, 41, 52, 56–57, 66, 82–83, 85, 88, 105, 108, 119, 146, 148
  and failures, 105, 107, 110–111, 121, 113

filter of, 36, 56, 91
and hazard, 37, 53, 55, 63, 75, 78, 144, 151
of hazardous events, 39, 61, 63, 147–148
indicators, 37
individual, 148
inference of, 66, 88
from interactions, 52, 85, 87, 149
of interfaces, 53
in interventions, 51–52, 54, 67–68, 84
in lifecycle, 36
in maintenance, 54
management, 24, 33, 35, 88, 128
occupational, 96
perceived, 116–117, 148
physical, 55
of places, 6
as a process variable, 64, 129
promoter, 86
as a random variable, 5, 51, 65–66, 87, 132, 147
reduction, 68, 150
responsibility for, 52–53
and safety, 150
and safety function, 37
sensor, 65, 67–68, 85, 89, 119
series of, 81
situations, 19, 60, 121
social, 148
tolerable, 33, 35–36, 51, 54, 64, 66–68, 151
in transportation, 34–35
treating of, 35
and value analysis, 93–94, 96, 102
Ritual, 4, 17, 20, 53, 70
Rule, 8, 21, 22–23, 25, 29–30, 34, 40, 67, 98, 110, 113, 117, 119–120, 122

S

Sabotage, 39, 111–112
Safety
  in abnormal occurrences, 128
  audit, 37
  barriers, 63
  and beliefs, 9
  concepts, 4, 38, 127, 133
  diagnosis, 7–8, 38
  "first", 11
  and fondness, 9
  function, 21, 37, 54, 89, 93, 99, 102–103, 150–151
  holistic approach/view, 11, 57, 150
  integration with vital functions, 11, 23, 103
  and leadership, 19, 114
  in the life cycles, 24
  management, 24, 38
  meetings, 7, 34
  methods/techniques, 151

Safety (cont.)
  monitoring, 8
    needs, 14, 23, 25, 96
    at organizations, 21, 40, 42
  performance, 21
  place of, 44
  procedures, 102
  professionals/experts/advisers, 34, 52
  programs, 24, 34, 36
  and reductionist view, 3
  and reliability, 150
  responsibility for, 24
  and risk, 150
  rules/standards, 21, 38, 113, 119, 129
  state, 7
  strategy, 89
  studies, 4, 93, 128–129
  and subjective view, 3
  subjects/matter/aspects, 34, 87
  team/group, 63, 127
  and value analysis, 42, 93, 150
  and values, 7
  at work, 27
Scenario, 27, 29, 39–40, 42, 45, 67, 148
Self-assertion, 2, 12, 23
Self-realization, 14, 23, 96
Sensitivity, 37
Seveso, 75
Shareholder, see sponsor
Shutdown, 5–6, 12, 76, 129, 142, 145, 149–150
Signage, 3–4, 48, 139
  and order-cleanliness, 8, 36, 57, 85–86
Signal, 4, 109–110, 123, 149
Skill, 6, 19, 27, 40–41, 43, 64, 68, 87–88, 107, 109, 111, 122
Society, 1, 29, 42, 67, 120
Sponsor, 11, 23, 42, 96
Standard Operating Procedure (SOP), 74–75, 113
Steinitz, W., 18
Strain, 110
Stress, 3, 109, 122, 143
  mechanical, 99, 108–109
Supervisors, 8, 34, 52, 91, 107, 111, 116, 119, 123–124, 102
Survival, see also happiness; perpetuation
  of organizations, 2, 23, 101
Susceptibility, 60, 69, 124, 127, 133–134, 140, 144, 146, 148
Synergy, 3, 11, 68, 102
System, 2–5, 51–3, 55–56, 71, 72–68, 83–87, 90–91, 99, 101, 103, 105–106, 108–109, 111, 114, 118–120, 123–124, 128–129, 134, 150
  emergency control, 36, 39–40, 47, 51, 56, 59–60, 81–82, 85–86, 89–90, 127–128, 133, 136, 139, 142–144, 146, 148
  ecological, 37, 129

  immunologic, 143
  living, 5–2, 12–13, 43, 51, 52, 129
  management, 6–8, 11–12, 20, 22–24, 27, 29–31, 33, 37–38, 56, 70, 73, 87–89, 110
  mechanical, 2, 37, 52, 109, 129
  operational, 5, 7, 12, 22, 56, 70, 124, 127, 129, 146, 149–150
  organizational, 6–7, 12, 22, 33–35, 40, 43, 56, 69–70, 146
  risk control, 7, 33, 52–53, 56, 60–61, 64–69, 79, 86, 89, 127, 139–145, 147, 151
  social, 37, 119, 129

T

Tao, 2
Target, 4, 7, 41, 47, 51, 55, 57, 60, 63, 69, 71, 86, 89–90, 98–99, 102–104, 127, 132–136, 138–139, 141–146, 148
Teamwork, 9, 93–94
"The Bridge on River Kwai", 112–113
Theory
  chess, 19
  failures, 76, 88, 106
  of systems, 2
Training, 6, 18, 27, 33, 36, 40–43, 56, 64, 68–70, 73, 86, 90–91, 102, 106, 109–110, 120, 145

V

Value
  personal, cultural, 1, 4, 6–9, 20–21, 23, 30, 34, 36, 38, 87, 89, 111–114, 116, 119–120, 128, 130
Venom, 110, 134, 136, 138–139
Victim, 39, 44, 46, 49, 128, 132
Vision
  Cartesian, 3
  of future, 8, 17, 20, 25, 29–30, 89
  global, 8, 34
  holistic, 2, 66
Visit, 42, 44, 52, 88, 111, 123–124, 145, 147
Vulnerability, 60, 69, 127, 134, 136, 140, 144–146, 148

W

Waste, 36, 51, 129
What-If, 57, 67, 77–78, 80, 82, 85, 88
Work, see also safety
  abnormal occurrences at, 67, 83
  absence from, 4, 6
  accident from, 8, 21, 33
  aggressive agent at, 138–139, 141
  carelessness at, 91, 122–123

conditions of, 66, 85-2, 108–110, 121–122
conscious failure at, 111–112, 122, 124
good practices at, 34, 57, 62, 69–70, 109–110
hazardous, 145
at home, 132
incapacity to, 61
intruder at, 91, 111
and lifecycle, 36
low load at, 147
mental, 1
needs at, 13–14
permit to, 36, 54
place of, 87, 109
repetitive, 6, 128
risk analysis of, 19
risk control at, 53, 69, 85
risk management outside of, 35–36
technical failure at, 122
unsafe behavior at, 115–116
World War II, 93, 110

Y

Yin, Yang, 2–3, 12

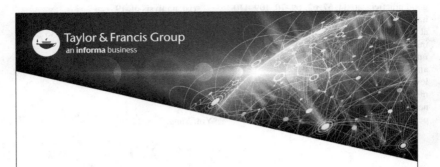